Bones of Contention

BONES of CONTENTION

Controversies in the Search for Human Origins

SECOND EDITION

WITH A NEW AFTERWORD

ROGER LEWIN

The University of Chicago Press
Chicago and London

The University of Chicago Press, Chicago 60637
The University of Chicago Press, Ltd., London

Library of Congress Cataloging-in-Publication Data

Lewin, Roger.
    Bones of contention : controversies in the search for human
origins / Roger Lewin.—2nd ed., with a new afterword.
        p.   cm.
    Includes bibliographical references and index.
    ISBN 0-226-47651-0 (alk. paper)
        1. Human evolution.   2. Human beings—Origin.   I. Title.
GN281.L487 1997
599.93'8—dc21   97-972
                                                        CIP

Acknowledgments

A book such as this one is in many senses a collective effort, a project that would have had no chance of succeeding without the willing and generous cooperation of the paleoanthropology profession as a whole. I should therefore like to express my deep gratitude and indebtedness to all those in the profession whom over the past two years I have bothered and pestered with requests for interviews, copies of ancient papers and manuscripts, access to correspondence, and yet more interviews. Never once was I refused a request, even though at times it must have been troublesome, inconvenient, and even personally distressing. This universal generosity was the more remarkable because my stated task was not to extol the triumphs of the science, but to show how and why the profession has at times erred.

Paleoanthropology, like all sciences, is an activity done by people and is therefore prey to the same kinds of subjective interpretations and personal interests that influence other activities done by people, such as politics. No scientist likes to be shown to be less than scientific, and yet virtually everyone I talked to helped me do just that. My goal—and perhaps the profession's too—was to demonstrate that paleoanthropology is one of the oddest of sciences, because it touches on some of the most basic and sensitive of questions that humans ask of themselves—namely, where do we come from? and what is our place in the world? And yet, for all that, paleoanthropology still works as a science. The profession as a whole encouraged me and helped me in pursuit of that goal, but it goes without saying that whether the book succeeds or not rests with the author alone.

To name any individual for thanks is to risk omitting others who may have influenced the book in less direct but still important ways. Nevertheless, I will take that risk and list below those who during the past two years and earlier have helped me create a modest oral history of their science. The names are listed alphabetically:

Peter Andrews, Frank Brown, Bernard Campbell, Matt Cartmill, Thure Cerling, J. Desmond Clark, Basil Cooke, Yves Coppens, Garniss Curtis, Brent Dalrymple, Raymond and Mrs. Dart, Michael Day, Robert Drake, John Durant, Ian Findlater, Frank Fitch, Andrew Gleadow, Stephen Jay Gould, Michael Hammond, John Harris, Andrew Hill, F. Clark Howell, William Howells, Tony Hurford, Glynn Isaac, Donald Johanson, William Kimbel, Kamoya Kimeu, Misia Landau, Mary Leakey, Meave Leakey, Richard Leakey, G. Edward Lewis, Jerold Lowenstein, Ernst Mayr, Ian McDougall, Henry McHenry, Jack Miller, Ashley Montagu, Todd Olson, Charles Oxnard, David Pilbeam, Charles Reed, Vincent Sarich, Birgitte Senut, Pat Shipman, Charles Sibley, Elwyn Simons, Frank Spencer, Christopher Stringer, Shirley Strum, Phillip Tobias Russell Tuttle, Alan Walker, Sherwood Washburn, Tim White, Allan Wilson, Milford Wolpoff, Bernard Wood, Adrienne Zihlman, Lord Zuckerman.

It is perhaps invidious to single out any individuals to whom I am more indebted than to others, but again I feel impelled to take that risk, because there are some without whose time and encouragement the book would simply have been impossible. They are, again alphabetically, Don Johanson, Mary Leakey, Richard Leakey, and Tim White. In addition, David Pilbeam not only let me endlessly rake over what he sees as his past mistakes, but also inspired me to do the book in the first place and ensured that I was thinking about the right kinds of questions.

I am grateful to the following institutions for allowing me to search through records both ancient and modern: American Museum of Natural History (Osborn Library and archives); British Museum (Natural History); Institute of Human Origins; L. S. B. Leakey Foundation; National Museums of Kenya (Leakey archives).

Finally, in the decade since I wrote the first edition, paleoanthropoligists have continued generously to include me as a close spectator of their endeavor, despite what I wrote about them! It has been a decade of remarkable discoveries in many realms, and I am privileged to have had the opportunity to be a commentator and critic of these developments, in places such as the journal *Science*, in books, and various popular magazines.

The untimely death of Allan Wilson in 1991 robbed paleoanthropology of one of the more creative minds in the business. His work in the 1960s and 1970s on the date of origin of the human family helped establish molecular anthropology as a sound discipline, even though his conclusions were at first strongly resisted by the anthropological community, as described in chapter 6.

And in the 1980s and early 1990s, he applied the same zeal and innovative thinking to the problem of the origin of modern humans. Once again, controversy, as discussed in the Afterword, surrounded his approach and conclusions, and persists beyond his death. The burgeoning volume of molecular work that is now such an important part of paleoanthropology is testimony to the fertility of the molecular approach Wilson pioneered.

To the memory of Allan Wilson

Contents

CHAPTER 1

BONES OF CONTENTION

Richard Leakey looked uncharacteristically tense. Famous son of that superstar family of African prehistory, Leakey was well used to appearing in public. Indeed, he is universally acknowledged as a gifted speaker, in both formal presentations and more casual gatherings, just like his father, the late Louis S. B. Leakey. But on this occasion Richard looked lost for words, groping for a response.

The occasion was an edition of *Cronkite's Universe*, a television series hosted by America's favorite avuncular figure, Walter Cronkite, for years the anchorman on CBS's nightly news program. On his *Universe*, Cronkite explores some of the scientific issues that interest him. Human origins—fossil man—interest him, so early in 1981 he thought he would ask Leakey to join him in a specially constructed studio in the bowels of the American Museum of Natural History, on the west side of New York's Central Park. The backdrop to the set was a large array of ape skulls—blank-eyed chimpanzees and gorillas staring into the camera. A small table at the front of the set carried more apelike skulls, but these were fiberglass casts of ancient fossils, property of Donald Johanson, who was Cronkite's other guest on the program.

"We brought Leakey and Johanson together in the American Museum of Natural History to discuss their different ideas about man's ancestry,"[1] explains Cronkite. Johanson, he reminds his viewers, has in recent years enjoyed the outstanding success of discovering a now-famous 3-million-year-old skeleton, nicknamed Lucy, and the remains of perhaps thirteen individuals of a similar era, which became known as "the First Family." Cronkite does not exaggerate when he describes the discoveries as "the century's most important fossil finds," all of which came from a remote spot in the Afar region of Ethiopia and are truly remarkable.

"Before the discovery of Lucy made Donald Johanson a celebrity, the king of the mountain of paleoanthropology was Richard Leakey," Cronkite continues, setting the scene for the TV event his viewers are about to witness. "Operating from his base camp at Lake Turkana in northern Kenya, Leakey has steadily unearthed numerous fossils. There was a two-million-year-old skull called 1470—that's a museum number—which brought Richard fame. He was credited with finding the oldest ancestor of man—until Lucy came along."

The scene is now clear: the viewers are to be shown a debate, a scientific confrontation between two of anthropology's most visible protagonists.

"There has been a controversy that has been going on now for nearly three years between Richard and myself, and it specifically focuses on the family tree," says Johanson. "We presented our family tree, let's see, it must have been in January 1979, and very shortly thereafter I know that Richard and others, but specifically Richard, had said that it does not fit the evidence of the fossil record."

Cut to Leakey.

"I've heard it all before. I think it is marvelous what you've done, Don. I just don't agree." Apparently surprised at the path the program is taking, Leakey attempts haltingly to head off a direct debate, his usual composure having departed. "I am not . . . I'm not willing to discuss specifics of why I think a bone means this or doesn't mean this . . . I've been around thirty-five years in a family that has seen lots of controversy. I've seen fossils in favor, out of favor, back in favor. Let us be . . . Let us stand back from it. Of course it is important, Don. I wouldn't minimize it. But I'm not going to say whether you are right or wrong." A short pause; he tosses back his head—a characteristic Leakey gesture—laughs, and ends, "But I think you're wrong."

The *Universe* program was being recorded in the spring of 1981.

Leakey was on one of his frequent but brief visits to New York, this one for a board meeting of the Foundation for Research into the Origin of Man, or FROM, an organization he had founded with the purpose of raising research money for anthropology. The previous two years had been grueling for Leakey, not least because in the fall of 1979 he had received a lifesaving kidney transplant from his younger brother Philip. He was embroiled in a messy reorganization of Kenya's National Museum in Nairobi, of which he is director, and the associated research institute that had been established in 1977 in honor of Louis Leakey. And he had completed a seven-part television series for the BBC, titled *The Making of Mankind*, which had involved extensive travel through four continents.

Johanson had been busy too. With fieldwork in Ethiopia on hold because of political problems, the young Cleveland anthropologist had concentrated his efforts on analysis of the unique set of fossils in his charge. He also displayed the kind of infectious energy and drive that has for so long been a hallmark of the Leakeys, both father and son, and set up his own research center, the Institute of Human Origins, in Berkeley, California. With numerous appearances on television as host of a science program, and many radio "phone-ins," Johanson was indeed rapidly achieving the celebrity status alluded to by Cronkite. He may not have been featured on the front cover of *Time* magazine, as Leakey was in 1977, but he was fast becoming America's best-known anthropologist.

Perhaps the rivalry between the two men was inevitable, especially in a discipline that seems to invite individualism and publicity. Perhaps there simply cannot be two "kings of the mountain of paleoanthropology," as Cronkite had put it. In any case, the differences of opinion between Johanson and Leakey had become clearly visible to the public eye, incomparably more so than any similar dispute in, say, an obscure side alley of entomology or even the more fashionable molecular biology. At one point a photograph of the two had appeared on the front page of *The New York Times* under the headline RIVAL ANTHROPOLOGISTS DIVIDE ON 'PRE-HUMAN' FIND." As on several similar occasions, the story in the *Times* reported that while Johanson was eager to explore the differences of opinion in the public arena, Leakey showed a distinct reluctance to do so.

Meanwhile, both men had written popular books. Leakey's was a wide-ranging volume on the origins of humanity and culture, which was a companion to the BBC television series. And Johanson's was a much more personal affair, entitled *Lucy*, which describes in a lively text his explorations in Ethiopia and the subsequent scientific conclusions. Both books received praise and criticism. One reviewer accused Leakey of virtually ignoring Johanson's work—"the single most important find in the last 20 years"—while another chastised Johanson for indulging in gossip, personal attack and innuendo, particularly at the expense of Richard Leakey.[2]

No wonder, then, that when Leakey was called from his FROM board meeting for a telephone call from his publisher urging him to appear on *Cronkite's Universe* with Johanson, his initial reaction was No, thank you. The publisher persisted, saying that the publicity would benefit sales of *The Making of Mankind*. Leakey asked what form the program would take, and was told that there would be no confrontation about Lucy, that both he and Johanson would be interviewed about human evolution and creationism, which was much in the news at the time. Leakey finally agreed, excused himself from the board meeting, and took a cab across Central Park to the museum, where Johanson and Cronkite were already waiting.

The two anthropologists shook hands, and Leakey asked if Johanson knew what was going to happen. "No; I've just been asked to come along as well." Leakey insisted that he didn't want to get into a debate about "Lucy and our alleged differences." Johanson said he thought they were there to talk about evolution and fundamentalism. Nevertheless, Leakey was apprehensive. "I was very uncomfortable," [3] he recalls. "I was in two minds to leave there and then, because I felt the whole thing was not right and not what I'd come to do." By this time, however, all three men were on the set, and were soon sitting around a small table. Leakey's disquiet sharpened when Cronkite asked if he had brought any props: "fossils or casts or anything?" Unable to oblige, having come unprepared and straight from a board meeting, Leakey simply replied, "No; I didn't bring anything with me." Johanson, however, was better equipped and produced a reconstruction of a Lucy-type skull, together with some other items. The interview started.

Instead of general questions about evolution and fundamentalism, discussion quickly turns to Johanson's and Leakey's differing views of the human family tree. "I got angrier and angrier," recalls Leakey, "but felt that once the cameras were running I couldn't walk out." His laughter on saying to Johanson, "I think you're wrong" was a sign of tension in the face of a situation over which he had no control—a very uncharacteristic position for a Leakey to be in.

Johanson's response is to say that, nevertheless, "it would be fun to show a portrayal of how I see the family tree."[4] Whereupon he leans to the side of his chair and pulls out a previously concealed chart that carries a very neat sketch of Johanson's view of human ancestry. It shows a simple Y-shape, with the common ancestor at the fork, the human line leading to *Homo sapiens* on one side and the now-extinct ape-men, or australopithecines, on the other. The common ancestor is the species known as *Australopithecus afarensis,* which is the name given to it in 1978 by Johanson and to which Lucy belongs.

Next to Johanson's tree is a blank space, which he invites Leakey to fill in with his version. "No, no, no . . . I haven't got crayons . . . I haven't got cutouts . . . I'm not an artist . . . I don't think I can do this." To oblige, Johanson hands Leakey a thick felt-tip pen. Leakey is left looking somewhat blankly at the chart, and inwardly declaring, "You bloody fool, Leakey. You walked right into this. You are caught. What are you going to do now?"[5]

Meanwhile, Johanson turns to the camera to begin an explanation, in the absence of a visual representation of the opposing view, of what in fact the differences of opinion are. Leakey interrupts, asks Johanson to hold one edge of the chart, and says, "I think I would probably do that," and he places a large X through Johanson's carefully drawn tree. "Well, what would you draw in its place?" challenges Johanson, who is visibly taken aback by Leakey's action. Regaining his composure somewhat, Leakey says, "A question mark," and does so boldly, filling his allotted space, and accompanying the gesture with another burst of Leakey laughter, this time more relaxed.

Johanson hurriedly drops the chart out of sight behind his chair, with Cronkite clearly enjoying the events enormously. Here was real television: no stuffy scientific "talking heads" for *Cronkite's Universe!*

In the absence of a hoped-for dialogue with Leakey, Johanson goes on to outline the differences that mark the two views. "To put it succinctly, Richard and his parents, Louis and Mary, have held to a view of human origins for nearly half a century now that the line of true man, the line of *Homo*—large brain, tool making and so on—has a separate ancestry that goes back millions and millions of years. And the ape-man, *Australopithecus*, has nothing to do with human ancestry."[6] The discovery of Lucy and her fellows, he goes on to suggest, shows that this argument is wrong, that the origin of the *Homo* line is recent, and that *Australopithecus* is right there in the middle of it all, our direct ancestor.

Leakey does not respond, but says that like Johanson, he would like to see a lot more fossils discovered, wherever they come from —Ethiopia, Kenya, Tanzania. "I would love to prove him right," Leakey says. Pause. "But I might just prove him wrong." Theme music rises. Credits roll. The confrontation is over—at least the public version of it.

Cronkite and the two anthropologists leave the set. Leakey gives Johanson a copy of *The Making of Mankind* and asks for a copy of *Lucy* in return. Coolly, the two part, with Leakey pursued by a producer's assistant who wants his signature on a broadcast release form. Leakey declines, and departs.

Persisting in his view that such public confrontations between the so-called "Leakey line" and Johanson's are contrived and scientifically unhelpful, Leakey now describes the Cronkite show as "Unfortunate."[7] Johanson's assessment is equally pithy, but more emphatic: "I won!"[8]

Controversy is no stranger to science, whatever the object of its study. Indeed, science advances by the repeated overthrow of established, accepted notions by new ones, which in their turn are later modified or thrown out. Science thrives on the progressive elimination of error, the continual updating of knowledge— which, by its very nature, is always provisional. And the updating process is frequently accompanied by vigorous sparring between proponents of the old and the new, whether it be in the marble halls of Victorian museums or amid the high technology of laboratories of molecular biology. After all, no one likes to be told that the notions upon which he may have built and promoted his career have turned out to be wrong. And scientists, contrary to the myth that they themselves publicly promulgate, are emotional human beings who carry a generous dose of subjectivity with them into the supposedly "objective search for The Truth."

In fact, a completely unbiased, unprejudiced exploration of nature is a methodological impossibility, as biologist and philosopher of science Sir Peter Medawar is fond of pointing out. Without a set of expectations to act as a guide, such a search would be a chaotic and largely unprofitable enterprise. Moreover, the way in which scientists typically report their findings, in formal papers submitted to learned journals, is, he says, "notorious for misrepresenting the process of thought that led to whatever discoveries they describe."[9] Preconceptions are rarely acknowledged, because this, after all, would be "unscientific." And yet preconceptions are an individual scientist's guide to how to view the world with a degree of order that allows structured questions to be asked.

The anonymous aphorism "I wouldn't have seen it if I hadn't believed it" is a continuing truth in science. And of course, it cuts two ways: you often see what you expect to see and not what you don't. Of course, no two scientists' sets of guidelines, or preconceptions, are going to be identical, even if the individuals concerned are in broad agreement. And as preconceptions are the lens through which each scientist views the questions to be asked about the world and the "facts" perceived therein, there is always a good deal of room for lively disagreement.

Johanson readily agrees that paleoanthropology is no different from other sciences in this respect. "The fossil finders themselves have often brought with them their own personal prejudices and beliefs . . . We see discoveries as bolstering our specific interpretation of what the family tree should look like."[10] Leakey's view is similar. "In our family we were working with the human sciences, and I was never shown examples of objectivity in the true sense of what science is supposed to be like."[11]

On the face of it, therefore, paleoanthropology appears to be little different from any other science in this respect, and its practitioners readily record the fact that disagreement about new fossils is, and always has been, rife. "Practically all paleontological discoveries can be described as bones of contention,"[12] says British anthropologist John Napier. He speaks from experience, having been embroiled in one of the livelier bone fights of recent times when, together with Louis Leakey and Phillip Tobias, he named a new species of hominid, in the mid 1960s, *Homo habilis.* Reverberations of the dispute continue to this day. "Almost every new discovery has started afresh such disputes as followed the finding of the Neanderthal skull,"[13] wrote Sir Grafton Elliot Smith, whose name is tied to the infamous Piltdown controversy. "Every discovery of a fossil relic which appears to throw light on connecting links in man's ancestry always has, and always will, arouse controversy," opined Sir Wilfred Le Gros Clark, the prominent British anthropologist, when he delivered the Huxley Memorial Lecture in 1958. Incidentally, he titled his lecture "Bones of Contention," which, for so courteous and proper an Englishman, was a strong statement. Moreover, his public disagreements over the shape of the human family tree with his Oxford colleague Solly (now Lord) Zuckerman and with Louis Leakey are prominent in the annals of the science of paleoanthropology, as is his involvement in unmasking the Piltdown fraud. And so it goes.

The paleoanthropology literature is replete with references of this kind to controversy, disagreements, and even personal battles. So when American anthropologist Ales Hrdlicka asked in 1927, "What is the actual, precise, evidence for human evolution that *science* now possesses, and upon which *it* bases far-reaching conclusions?" (my emphasis), [14] he was in fact posing a question that has no answer. Not because there is no evidence for human evolution, but because no science works that way. No science—least of all paleoanthropology—is as objective as Hrdlicka implies here or as is often portrayed in the philosophers' idealized view of science.

Paleoanthropology is thus no different from other sciences in being controversial. What sets it off from other sciences is the *degree* of controversy it engenders. Yes, controversy is found in all sciences, but in paleoanthropology discernibly more so. Yes, preconceived ideas shape the progress of all sciences, but nowhere else to the degree that occurs in the search for human origins. And yes, personalities are important in the flow of all sciences, but, again, in the science of man emphatically so. "All sciences are odd in some way," notes Duke University anthropologist Matt Cartmill, "but paleoanthropology is one of the oddest." [15] Paleoanthropology is like all other sciences, only more so. Why?

Why, when the fossil missing links under consideration are those in a chain of extinct horses or ammonites, is there controlled controversy—"But when it extends to fossils which can be brought forward as evidence that man is related to something simian," observes Gerrit Miller, onetime curator at the National Museum of Natural History, Washington, D.C., "the case is very different." "Then," he notes, "it leads to the expressing of opinions delivered from sharply defined and diametrically opposed points of view." Why?

Le Gros Clark has an answer: "Undoubtedly, one of the main factors responsible for the frequency with which polemics enters into controversies on matters of paleoanthropology is purely an emotional one. It is a fact (which it were well to recognize) that it is extraordinarily difficult to view with complete objectivity the evidence for our own evolutionary origin, no doubt because the problem is such a very personal one." Ernst Mayr, one of this generation's most prominent evolutionary biologists, concurs: "Human beings seem quite incapable of speaking about themselves and their history without becoming emotional in one way or another."

Observe what happened when, at the beginning of 1984, the American Museum of Natural History in New York held an unprecedented exhibition of original fossils relating to human origins. The so-called *Ancestors* exhibition was years in the planning and deep in the agonizing, because it involved persuading museum curators the world over to part briefly with their priceless and fragile relics, which would be shipped to New York and placed on public view, some for the first time ever. Curators arrived at John F. Kennedy Airport, having carefully cradled their charges in first-class seats, to be met by a motorcade of limousines and police escort. No VIP had more attentive care and reception.

Before the forty or so precious fossils went on public display, protected by bulletproof glass, they were the focus of one of the most extraordinary anthropology workshops of all time. In a small room on the museum's second floor a dozen anthropologists at a time gathered to examine the ancient skulls and limb bones, comparing one with another in a way that had never previously been possible. Excitement was high, but voices were often hushed, as long-held differences of opinion were aired over the objects of disagreement. "It was like discussing theology in a cathedral," said Michael Day, a British anatomist and longtime associate of the Leakey family. "To be in the same room with all these relics was for many workers an emotional event," agreed Christopher Stringer, an anthropologist at the British Museum (Natural History). "Sounds like ancestor worship to me" was the comment of a sociologist of science who was observing the events. It is hard to imagine a group of, say, biochemists becoming quite so emotional in the presence of their favorite experimental organism, *Escherichia coli*.

There *is* a difference. There *is* something inexpressibly moving about cradling in one's hands a cranium drawn from one's own ancestry.

The *Ancestors* exhibition displayed yet another realm in which paleoanthropology is different from most other sciences. A significant number of fossils that the organizers hoped would be present did not arrive, for a variety of reasons—some overtly political, some more subtly so. Lucy and her fellows, for instance, were absent because the Ethiopian authorities were in the process of establishing ground rules for handling antiquities research and material, particularly where foreign nationals were concerned. Some of the Leakeys' material from Olduvai Gorge failed to appear because at the last minute Tanzanian authorities objected to South African participation in the exhibition. The Chinese declined, again at the last minute, to send their famous Peking Man fossils —possibly because of memories of the loss of many of them when they were to be shipped to the United States for safekeeping at the beginning of the Second World War, and perhaps because damage had occurred to a fossil that had recently been lent to another international exhibition. Nothing arrived from Australia because the aborigines had recently begun to object to the general way in which the relics of their ancestors had been treated by the colonial scientists through the years. Kenya declined to ship any fossils on the ground that the risk of damage was too great. The absence of Richard Leakey from the workshop was as notable as the absence of his fossils. And so it went.

Politics rarely impinge overtly on scientific interchange, but here it did in a most explicit manner. No one was in any doubt that the nature of the subject at hand—the origin of mankind— exacerbated the political sensitivities that were being asserted.

If there is an emotional significance to the fossils themselves, so too is there in their discovery. "Not every expedition is like *Raiders of the Lost Ark*," says Johanson, "but they have their moments."[17] After some years of being denied access to the fossil sites in Ethiopia because of political problems there, Johanson recently organized an expedition to Olduvai Gorge, the site of much of Louis and Mary Leakey's work. "We were pleased to have an opportunity to work in the field again, to do the sorts of things we've been trained to do, the sorts of things we really like doing. We like banging around in vehicles, looking for fossils, wandering around in the sun. It's a heck of a lot of fun. It's what the business is all about."[18]

Johanson recalls the thrill in 1973 of finding his first hominid fossil in Ethiopia, a 3-million-year-old knee joint. "A moment of magic." And he acknowledges that the search is often spiced by hopes that are not always strictly scientific. "We have a passion to find the oldest, the most complete, the biggest-brained, the most enigmatic fossil," he recently told an audience at a public lecture at the American Museum of Natural History in New York.[19] Many anthropologists feel like this, but few are candid enough to express it publicly.

Given the emotional content of the search and the thrill of the find, it is perhaps not surprising that practitioners sometimes become possessive about "their" site and "their" fossils. One of the most extravagant examples was that of Eugène Dubois, who unearthed some of the first fossil human ancestors to be discovered, *Pithecanthropus erectus*, in Java in the early 1890s. One observer commented: "*Pithecanthropus* became Dubois's destiny. It was his discovery, his creation, his exclusive possession; on this point he was as unaccountable as a jealous lover. Anyone who disagreed with his interpretations of *Pithecanthropus* was his personal enemy." Few since Dubois's time have gone to such extremes, which included secreting the fossils beneath his dining room floorboards; but elements of his attitude have always been present, and probably always will be.

The very fact that Johanson recently chose to go to Olduvai Gorge at a time when his relations with the Leakey family are at best described as strained has itself been the cause of a good deal of emotionalism. Although the Leakeys don't have any special rights to Olduvai beyond a long association with the site, and Mary Leakey no longer carries out fieldwork there, many in the profession considered it was "bad form" for Johanson to "trespass" as he did. Such is the nature of paleoanthropology.

With a limited number of fossil sites available to work, and a still pitifully small inventory of fossils to analyze, all of which may be in the control of just a few people, research access has always been a sensitive issue. Sensitive not necessarily because anyone has in fact been excluded for inappropriate reasons, but because in the emotionally charged atmosphere that sometimes prevails in paleoanthropology, there is always the possibility that the charge of inappropriate exclusion will be raised. In any event, this issue has frequently been a component of the many controversies that punctuate the progress of the science, and often a disruptive one at that. "Sometimes this has resulted in rather bitter rivalries," says Johanson, "with scientists breaking down communication with one another. . . . This is unfortunate, because it stops the development of the science. It interjects a distasteful form of elitism, because it sometimes results in instances where —it has been recently written—only those in the inner circle get to see the fossils; only those who agree with the particular interpretation of a particular investigator are allowed to see the fossils."

Virtually every anthropologist has a tale or two to tell about a rival professional improperly preventing others from working on fossils in his possession. "There are lots of ways of simply making it difficult for someone to come to your lab and work with the fossils, if you choose not to have them come," comments one senior anthropologist. "You don't have to be so obvious and crude as to say 'No,' even if that's what you really intend." Of course, even when a curator of fossils has genuine reasons for suggesting to a fellow anthropologist a more convenient time to come to his lab, for example, or for imposing some kind of restriction on publication, such responses can easily be misinterpreted as malicious attempts to prevent access, and not infrequently they are.

Of course, the development of techniques for producing superb casts, which can be distributed to many laboratories, is easing this problem of restricted access. Nevertheless, there is the strict protocol to be followed that the discoverer of a fossil have the first opportunity to describe it formally and to analyze it.

Naming a new species, however, is a different matter. In principle, anyone can do that. "Yes, anyone can name a new species, not just the discoverer,"[20] confirms Johanson. And there have been some wonderfully notorious cases in the past. Perhaps the most outrageous were the great forays into the American Midwest by Edward Drinker Cope and Othaniel C. Marsh, two nineteenth-century paleontologists who went separately in search of dinosaur bones and other fossil relics. So keen was the desire of each to score a point against the other that they would hire people to buy fossils from the sources of their competitor. And such was the race to name new species before the other did that they tried telegraphing the appropriate messages back East, often with the most hilariously garbled result. Paleoanthropologists have not as yet tried to scoop each other with quite the bravado displayed by Cope and Marsh. But there have been some delicate moments.

So although Johanson acknowledges that there is no absolute right of a discoverer to also be the namer of the species, he says that there's a question of etiquette. In any case, the naming of a new species can be seen as a just reward. "All of us who are involved in the exploration business have to sacrifice something when we do it, whether it is security, or the comfort of being at home, injury and so on. With that sort of adventurism is a sense of wanting a reward. Part of that reward is discovery, and you'd like to be recognized for that. In this science one of the ways you have an opportunity of being recognized is if you find something new and different, you can name a species."[21] Whenever there is formal citation in the literature, the name of the species is followed by its author: for example, *Australopithecus afarensis*, Johanson, 1978, is the species to which Lucy belongs. And the rules of zoological nomenclature are such that once assigned, the species name and the associated author are virtually inscribed in stone.

Johanson is quick to point out that the naming of a species is often only the starting point of the really important work, which is the analysis of its significance. But is is easy to appreciate the attraction of that little bit of immortality which attaches to the naming of a species, especially when the species is identified as one's ancestor. This distinctly nonscientific frisson of paleoanthropology again sets it apart from other sciences.

In most fields of science the great names are typically people who have wrought some kind of significant intellectual advance —a new concept or theory. Most people have heard about Einstein's theory of relativity, but very few understand what it means in practical, tangible terms. With paleoanthropology it is different. It is the tangible discoveries that bring fame, not the intellectual theories. Many people have heard of skull 1470 or Lucy, but few would be able to say how they affect the theories of human evolution. Michael Day, a British anatomist and colleague of both Louis and Richard Leakey, identifies this as an unusual but important element in paleoanthropology. "Nine-tenths of your importance in this field comes from your finds. People remember Dart. They remember Johanson. They remember the Leakeys. But Le Gros Clark, who was an intellectual in the discipline, his name will fade. So will Clark Howell's. And certainly so will mine, because I'm not a finder. There is a tremendous bias towards finders. And with this goes an unwarranted weight on their opinions." Day is stoical enough to add, "I can easily be accused of sour grapes, of course."[22]

With the emotional elements of adventure, sacrifice and reward all compounded in the discovery of a new fossil, together with the soul-stirring aspects of ancestor worship, it might seem that the odds are heavily stacked against an objective analysis by the individual who by custom has the right of first pronouncement. Earnest Hooton, a prominent Harvard anthropologist of the 1930s and '40s, recognized this trap as "the psychology of the individual discoverer and describer." He wrote that "The tendency towards aggrandizement of a rare or unique specimen on the part of its finder or the person to whom its initial scientific description has been entrusted, springs naturally from human egoism and is almost ineradicable."[23] Compared with the number of practitioners in the field, the fossils available for analysis are relatively few, he said. The individual lucky enough to have first access to a particular specimen is therefore likely to "leave no bone unturned in his effort to find new and striking peculiarities which he can interpret functionally or genealogically. Unless he is very experienced, he is prone to discover new features which are partially the creations of his own concentrated imagination."

But Hooton identified an even greater danger. This is "the psychological conflict in which the discoverer or describer is torn between his desire to find primitive, unique, or anthropoidal features which will allow him to place his specimen nearer to the apes than any previously recorded, and his equally powerful urge to demonstrate the direct and central position of his new type in the ancestry of modern man." When the former impulse is in the ascendancy, says Hooton, "the author is likely to blow the dust off his Greek and Latin dictionaries and perpetrate some horrid neologism in creating a new zoological species, genus or even family, thereby committing simultaneously mortal sins in both philology and taxonomy." When the latter impulse succeeds, the describer "may seize upon metrically or morphologically insignificant features common to both [modern man and the fossil under study] as evidence of their genetic relationship." In other words, on the one hand you exaggerate the differences between your fossil and modern humans, thus getting for yourself a nice, ancient, discrete ancestor. And on the other, you overlook the differences and exaggerate the similarities, thus setting your fossil on the threshold of the noble *Homo sapiens*. The reader will see many examples of inherently anthropocentric interpretations of fossils in later pages of this book.

If all this were not bad enough, Hooton warns that "in addition to the frailties inseparable from the enactment of the role of original describer, one must also discount the author's previous commitments on the subject of fossil man, the ghosts of earlier opinions which rise to haunt him in the interpretation of new evidence." A dispassionate analysis of new fossil evidence is possible, he says, "only when one awaits the reworking of the material by persons not emotionally identified with the specimen." Even then, an independent analyst, while not potentially blinded by emotional attachment to a fossil, will still have a particular set of preconceptions against which he will judge it. So dispassionate it may be, but totally objective it can never be.

Hooton was writing in 1937, when indeed there were very few fossils available and when the tendency to christen each new one as a new species was at a frenetic pitch. Ernst Mayr remembers with dismay this species-naming frenzy. "By the 1950s, the student of fossil man had to cope with 29 generic and more than 100 specific names, a totally bewildering diversity of types."[24] Le Gros Clark was equally distressed by the spectacle. "Probably nothing has done more to introduce confusion into the story of human evolution than the reckless propensity for inventing new (and sometimes unnecessarily complicated) names for fragmentary fossil relics that turn out eventually to belong to genera or species previously known." Instead of filling gaps in the story of human ancestry, this habit tended "to produce gaps that did not exist."[25]

It is an unfortunate truth that fossils do not emerge from the ground with labels already attached to them. And it is bad enough that much of the labeling was done in the name of egoism and a naive lack of appreciation of variation between individuals: each nuance in shape was taken to indicate a difference in type rather than natural variation within a population. This problem has in some part been eased in the half-century since Hooton made his pithy remarks. But it remains inescapably true that applying the correct label is astonishingly difficult, not least because such labels are in a sense arbitrary abstractions; and especially so when the material on which the analysis is being done is fragmentary and eroded. "It is an incredibly difficult problem," says Lord Zuckerman. "It is one so difficult that I think it would be legitimate to despair that one could ever turn it into a science."[26]

The number of species recognized today as being part of the human family tree is mercifully small: just half a dozen or so, the multitude of species names of the early decades of the century having been rationalized by individuals other than those who found the fossils. And the naming of new human ancestors is a now rare event too: just two major species in the past twenty-five years. The first was *Homo habilis,* named by Louis Leakey and his colleagues in 1964. And the second, Johanson's *Australopithecus afarensis,* christened in 1978. And as mentioned earlier, both events occasioned storms of protest within the anthropological community. Zuckerman was scathing about the furor surrounding *Homo habilis.* "The debate in the press seemed to me less like a scientific discussion than a public auction of anatomical specula-tions," he said.[27] And the naming of Lucy provoked such a flurry of arguments and public controversy that Walter Cronkite thought it all worthy of broadcasting to a national television audience, as we saw at the beginning of the chapter.

So it's clear that, as Johanson observes, "Controversy is still dominant in the field, and always will be."[28] As we've seen, this controversy transcends simple intellectual disagreement, even though such dispute is part of it. Primarily, this controversy springs from deep within the protagonists. It has to do with self-image and an intimate identification with the subject of the de-bate. In Zuckerman's words, "So much glamour still attaches to the theme of missing links, and to man's relationship with the animal world, that it may always be difficult to exorcise from the comparative study of Primates, living and fossil, the kind of myths which the unaided eye is able to conjure out of a well of wishful thinking."[29] Nevertheless, as Johanson so neatly puts it: "It is better to debate the problem without solving it than solve it with-out debating it."[30]

The nature of that debate—in truth, it is a series of debates—is what is explored in the following chapters. There are four simple themes in the paleoanthropological debates—themes that some-times are dominant in scientific discourse, sometimes fading into the background, depending on the flow of the moment. They are the Who?, Where?, When?, and Why? questions, just like the clas-sic opening paragraph to a newspaper story. Who was our ancestor? Where did it first arise? When did we break away from the rest of the animal world? And, Why did it happen?

It was the Who and When questions that Leakey and Johanson were addressing in their debate on *Cronkite's Universe.* For them, the Where? question is pretty much agreed: Africa. Just half a century earlier, two other great names in paleoanthropology,

Henry Fairfield Osborn and William King Gregory, were conduct-
ing a very similar debate in that same building, the American
Museum of Natural History. For Osborn and Gregory, the Who?
and the Where? questions were uppermost, and When? was of
much lesser concern. But the tone of the Osborn–Gregory debate
was just as spirited, just as publicly visible, despite the absence of
television, and—most important of all—just as much focused
through the lens of preconceptions, and as loaded with the sensitive self-image of humanity. And so it has been throughout the
history of the science—driven, perhaps, by that ultimate of all the
questions: Why?

Why did it happen?

CHAPTER 2

THE STORYTELLERS

M isia Landau was sitting in Yale University's Sterling Library, its leather-covered chairs and high book stacks imposing a palpable sense of Ivy League academia. It was the middle of her doctoral years, 1979, and she was reading intensely *The Morphology of a Folk Tale* by Vladimir Propp, a Russian literary critic. The occasional whispered exchange hung in the Gothic hush. Books placed on old oak made that curious muffled thud that happens only in such places. Something modern was being played on the university's famous carillon. But Landau was hardly aware of any of this. She was preparing to run to the anthropology book stacks. But, restrained by centuries of tradition, she walked quickly instead, her heart thumping with excitement. "When I got to the shelves, the titles leaped out at me: *The Story of Man . . . The Adventure of Humanity . . . Adventures with the Missing Link . . . Man Rises to Parnassus.* Looking at them, I knew I had made a discovery. It was like finding a fossil."[1] She had discovered a missing link between literature and paleoanthropology.

Only two years earlier Landau had been spending most of her time in a neurology laboratory, a conventional white-coated scientist. Having completed a human-biology degree at Oxford University, England, she had enrolled in the graduate anthropology program at Yale and was hoping to uncover something significant about the evolutionary history of the human brain, specifically the reason for its rapid expansion around 2 million years ago. In 1972, Richard Leakey had discovered what was thought to be a 3-million-year-old, large-brained human ancestor—code-named 1470—on the eastern shore of Lake Turkana in Kenya. So brain evolution was very much in the news when Landau enrolled at Yale. And as Yale was in its "hyperempiricist phase," with anthropologists shunning "soft" subjects like culture, preferring instead to do "hard" science, which involved measuring things, Landau thought she would learn something about human evolution by studying chicken brains.

It was a short-lived venture. "I quickly realized that I was going to be a mediocre laboratory scientist. And that the questions I was asking about human brain evolution were not going to be answered for a very long time—and not by me." She needed a new direction, but what would it be? Making such a change is always a depressing and confusing experience for a graduate student, especially when one project has been tried and failed. Landau talked with her thesis supervisor, David Pilbeam, a young British anthropologist who had quickly established himself as a leading figure in the science following his move to Yale in 1963. All that Landau knew was that she wanted to do "something theoretical." Pilbeam had long nurtured a strong interest in the history of ideas in paleoanthropology, and says that given his time again, he would be a historian rather than a prehistorian. The combination of Landau's inclination to do something theoretical and Pilbeam's historical perspective launched the dissertation in a new direction: it would be some kind of analysis of early paleoanthropological ideas.

However, a further ingredient was to be crucial in the new venture, though neither teacher nor student realized it at the time. That ingredient was literature, Landau's great love as a young girl. "Doing science was something different for me. But, as things turned out, this dissertation took me back to something I am good at, something I am sensitive to: literature."

Every paleoanthropologist is familiar with the great names of the 1920s and '30s—the Americans, Henry Fairfield Osborn, William King Gregory, Frederick Wood Jones, and the British school of Sir Arthur Keith, Sir Grafton Elliot Smith, Sir Arthur Smith Woodward—but few can claim more than a passing acquaintance with their words. Landau immersed herself in them. Each of these authorities was concerned with the same problem: the scientific explanation of human origins. And yet disagreement between them was common and sometimes profound. "It seemed to me that they were talking about quite different problems, not the same thing at all," she recalls. "Their world views were quite distinct from one another." This latter observation had prompted a British sociologist at Yale, Keith Hart, to suggest that Landau might find it useful to look at some French structuralism and Russian formalism. A far cry from paleoanthropology—but, given her passion for literature, Landau needed no more encouragement.

"I started reading this material, and couldn't stop. I started making connections between literature and the anthropology texts. I started thinking in terms of a plot in these books. It was very exciting." A friend lent her a copy of Propp's *Morphology of a Folk Tale*, which is a classic work in literary analysis. The book was intriguing to Landau from the start, "because the title—'Morphology of a folk tale'—seemed to indicate some kind of anatomical approach to literature, and I was beginning to take a literary approach to anatomy." On the basis mainly of Russian literature, Propp describes the hero myths of folk tales in terms of a basic structure they all follow: the hero enters; is challenged by, and overcomes, a series of tests; and finally triumphs. Propp was very systematic, and he broke this basic structure down into a sequence of separate functions; the precise identity of characters and activities at each point may differ, but the underlying structure remains the same.

The more Landau read, the more she perceived connections. "I was sitting there, in the Sterling Library, reading Propp, and the folk tales seemed so . . . familiar . . . I suddenly realized that the tale also described human evolution, at least as written about in the books I'd been reading." This was the point of discovery. When she got to the paleoanthropology shelves, she now recalls, "I realized that I was standing in front of a genre of literature, that I could approach the study of human evolution as a study of literature." In other words, while Osborn, Gregory, and their colleagues considered themselves to have written scientific analyses of human evolution, they had in fact been telling stories. Scientific stories, to be sure, but stories nevertheless.

The main events in these stories are four, and they represent the evolutionary transformation of some kind of primitive primate ancestor into a civilized human being. They are as follows: the shift from the trees to the ground—terrestriality; the change of posture from walking on four legs to balancing on two—bipedalism; the expansion of the brain, with flowering of intelligence and language—encephalization; and the emergence of technology, morals, and society—civilization. Osborn and his four most prominent contemporaries agreed emphatically on the need for these four components of our transformation, but equally emphatically disagreed on the order in which they occurred.

Osborn, for instance, perceived the sequence of events as given above. It began with a distant ancestor leaving the trees and beginning life on the ground—terrestriality. This was followed by the development of bipedality; then brain expansion; and last, civilization. Overall, this is very much how Darwin saw things. For Keith, it was different. He envisaged the evolution of bipedalism in a still-arboreal ape, and the adoption of a terrestrial existence came second. The expansion of the brain—encephalization—followed the elaboration of technology and society, and did not precede it as Osborn believed. Elliot Smith had yet a different view, which placed encephalization as the first event. His brainy ape then became bipedal, while still leading a mainly arboreal life. Only then did Elliot Smith's ancestor come to the ground, whereupon the development of civilization followed. Gregory's scheme involved terrestriality as the first event, followed by the evolution of society and technology; bipedality and encephalization concluded the story in that order. Wood Jones's ideas were similar to Elliot Smith's, in that the ancestor became bipedal and brainy while still living in the trees. The order of events was, however, different: for Wood Jones bipedality preceded encephalization.

Each author had his own reasons for casting the evolutionary scheme the way he did, but there is order in the apparent chaos, argues Landau, because all followed the same basic structure in their narratives: the form of the hero myth.

Where Propp identified thirty-one functions within the narrative of the hero myth, Landau simplifies the analysis to nine. These include the introduction of the humble hero (an ape, a monkey, or a diminutive prosimian) in an initially stable environment; our hero is then expelled from this safety (because of climatic change) and is forced to embark on a hazardous journey during which he must overcome a series of tests (new environmental conditions) and thereby display his worth (develop intelligence, bipedalism, etc.); thus endowed, our hero develops further advantages (tools, for Osborn; reason, for Keith), only to be tested again (the rigors of Ice Age Europe); the ultimate triumph is the achievement of humanity. "There is a final irony," says Landau. "Again and again we hear how a hero, having accomplished great deeds, succumbs to pride or hubris and is destroyed. In many narratives of human evolution there is a similar sense that man may be doomed, that although civilization evolved as a means of protecting man from nature, it is now his greatest threat."[2]

Certainly, Osborn and his contemporaries frequently expressed themselves in the language of epic tales. Keith, for example, is clearly talking about a hero when he writes: "Why, then, has evolutionary fate treated ape and man so differently? The one has been left in the obscurity of its native jungle, while the other has been given a glorious exodus leading to the domination of earth, sea, and sky."[3] Roy Chapman Andrews, Osborn's colleague at the American Museum of Natural History, New York, voiced a similar pioneer spirit: "Hurry has always been the tempo of human evolution. Hurry to get out of the primordial ape stage, to change body, brain, hands and feet faster than it had ever been done in the history of creation. Hurry on to the time when man could conquer the land and the sea and the air; when he could stand as Lord of all the Earth."[4]

Elliot Smith, writing of the early history of humanity, finds it ". . . well within the bounds of reasonable conjecture to picture the wide stretch of Southern Asia and Africa as peopled by weird caricatures of mankind, roaming far and wide to satisfy their appetites and avoid extinction. In this competition, the distinctive characters of Man were fashioned in the hard school of experience." Indeed, Elliot Smith is particularly fired by the drama of it all. He writes about ". . . the wonderful story of Man's journeyings towards his ultimate goal . . ." and ". . . Man's ceaseless struggle to achieve his destiny."[5]

Osborn himself wove especially stirring prose, displaying a tremendous belief in the adventure and theater of it all. Phrases such as "The prologue and the opening acts of the human drama . . ."[6] and "The great drama of the prehistory of man . . ." reveal a palpable pride in his hero and a distinctly literary portrayal of evolution. Osborn's hero was forced to immense efforts: "The struggle for existence was severe and evoked all the inventive and resourceful faculties and encouraged him to the fashioning and first use of wooden and then stone weapons for the chase . . . It compelled Dawn Man . . . to develop strength of limb to make long journeys on foot, strength of lungs for running, and quick vision and stealth for the chase."[7] And the consequences of eschewing a demanding existence and choosing instead an easy life were evolutionary oblivion: ". . . the rise of man is arrested or retrogressive in every region where the natural food supply is abundant and accessible without effort." Neanderthal Man came in for some of Osborn's moral scorn in this respect. He represents, Osborn said, a splendid example of arrested or even retrogressive development. The reason? "Game was very plentiful, the rivers of France and England abounded with hippopotami which afforded an easy source of food supply, and in the forests and plains roamed many types of elephants and rhinoceroses." Such easy living was no suitable setting for Osborn's hero, who was "the highly superior race of man called the 'Cro-Magnon.'[8] Edward Grant Conklin, a contemporary of Osborn's and professor of biology at Princeton University, expressed the sentiment very succinctly: ". . . the lesson of past evolution teaches that there can be no progress of any kind without struggle."[9]

Elliot Smith's hero, like Osborn's, is rewarded with success only through effort. Our ancestors, he said, ". . . were impelled to issue forth from their forests, and seek new sources of food and new surroundings on hill and plain, where they could obtain the sustenance they needed." The apes' inferior fate was in their own hands. "The other group, perhaps because they happened to be more favorably situated or attuned to their surroundings, living in a land of plenty, which encouraged indolence in habit and stagnation of effort and growth, were free from this glorious unrest, and remained apes, continuing to lead very much the same kind of life (as Gorillas and Chimpanzees) as their ancestors since the Miocene or even earlier times." To emphasize the point, he adds the following: "While man was evolved amidst the strife with adverse conditions, the ancestors of the Gorilla and Chimpanzee gave up the struggle for mental supremacy because they were satisfied with their circumstances."[10] In other words, says Duke University anthropologist Matt Cartmill, putting such sentiments in historical context: "Darwinian man is lord of the earth, not because of any God-given stewardship or romantic affinity to the World Spirit, but for the same good and legitimate reason that the British were rulers of Africa and India."[11] And, presumably, for the same good and legitimate reason that Osborn's America was the acknowledged pioneer of a new age.

The epic nature of much of this writing is evident from the tone of the language once one has been alerted to it. And perhaps this should not be too surprising, given that the main subject of the book was so very close to the author's own experience: *Homo sapiens*. Although experts on trilobites or sea slugs have frequently been known to wax anthropomorphic about their subjects of study, when an author's subject is in essence himself, an even greater degree of identification and purple prose can be expected: the author really does have a hero to write about. And there is a story to be told, too, a sequence of events leading from one thing to another—from apes to us—and so a narrative style seems to be invited. But the fact that these texts generally conform in many details of their structure to the classical hero myth had not been predicted and is therefore truly significant.

In all likelihood, these authors must have been repeatedly exposed to hero myths—hearing fairy tales throughout childhood, for instance. They probably passed them on to their own children by turn. These myths were, after all, part of their cultural milieu and still are today. But the fact that this genre of literature should be discovered to be embedded in their mature writing about serious subjects is a little shocking. Or is it?

"When I first read Landau's work, I became worried,"[12] admitted the late Glynn Isaac, a leading archeologist and paleoanthropologist at Harvard University. "Is there a common structure to the human mind as trained by participation in culture—such that it seeks 'narrative' as a satisfying form of explanation?" he wondered. "From the examples she presents, from one's own aroused self-awareness, and from one's general knowledge of paleontological and archeological writing, the answer clearly seems to be 'yes.' I had only been dimly aware of this."[13]

Having come to appreciate this, Isaac realized that there was an obvious challenge to paleoanthropology: "Was there any way to present a sequential account that did not involve the hero-story structure? If not, did that disqualify study of the narrative of human evolution from being a science?" Isaac posed these questions to scientists who had gathered at Darwin College, Cambridge, in April 1982 to commemorate the centenary of the great man's death. "I have got used to the idea now," he continued, "and would argue that provided the fit between the stories and empirical evidence is improvable through testing and falsification, then indeed it is science."[14]

As an aside, Isaac added: "If any of the rest of the scientific community is inclined to snigger at the embarrassment of paleo-anthropologists over all this, pause and reflect. I bet that the same basic findings would apply to the origin of mammals, or of flowering plants, or of life . . . or even the big bang and the cosmos." To judge from Landau's experience in presenting her thesis to mixed scientific audiences at many international meetings, Isaac is right. "People always come up to me after my talk and say, 'You should take a look at our science; I'm sure it's going on there too.' And this is from physicists, ecologists, even biochemists—all kinds of scientists."[15]

Reaction among other paleoanthropologists has been mixed. One of the first to read Landau's dissertation outside Yale was Sherwood Washburn of Berkeley, one of the science's leading figures. Landau was bold enough to thrust her thesis into his hands during the April 1981 annual meetings of physical anthropologists in Detroit. Washburn was not especially happy at having to lug the hefty document around with him, but began to read it on the flight back to California. "I quickly became fascinated by it," he recalls. "It is a very useful idea. It makes it much easier to change your ideas. Once you have, in quotes, a Scientific Theory, all capital letters, there is a great resistance to change."[16] As soon as he arrived home Washburn wrote a quick note to Landau saying that her thesis was ". . . a new and most useful way of approaching human evolution."[17] A month later, having had a chance to read more of the thesis, he wrote again: "A day ago I gave an evening lecture for our undergraduate anthropology club. I'm told the lecture was a great success, and discussions continued for one and a half hours after it was over. I used your ideas, with full credit to you. . . ."[18]

Even an enthusiast like Washburn, however, has some reservations, particularly about how valid the ideas are in the modern science. "The more sophisticated the science becomes, the less applicable is the thesis,"[19] he says. Landau came face to face with this sentiment, but expressed even more forcefully, when she gave a seminar in the anthropology department at Berkeley. "Don Johanson was adamant that even if people told stories in the past, they certainly didn't now. The science is now so sophisticated, so objective that he for one is engaged in an unbiased search for the truth,"[20] Landau reports. Johanson's opinion is not uncommon among modern practitioners, and may well be the majority view.

To which Landau replies: "Scientists are generally aware of the influence of theory on observation. Seldom do they recognize, however, that many scientific theories are essentially narratives."[21] It's true, she admits, that people don't any longer tell the grand stories the way Osborn, Elliot Smith, and others did. "That's because they want to be more scientific and because the fossils that are available now do have a constraining effect on theories. But there are elements in the current literature of what was there in the twenties and thirties."[22] Storytelling, she notes, is what makes us human.

In addition to the narrative tone and structure that are built into accounts of human origins, there are other basic assumptions too, which begin to pop out of the page when even a mild degree of attention is paid. They all make the story so much more compelling. One of them is the result of telling a story whose end is already known. Now, obviously, one cannot tell a story unless one knows what is going to happen. But the effect in stories of human origins is to make authors view each step in a chain of events as in some way a preparation for the next—the end product, necessarily, to be *Homo sapiens.* A second, somewhat related element is the idea of progress, that evolution is a program of constant improvement, whose crowning glory is us. And the third, again intertwined with the earlier ones, is the belief that man is the inevitable outcome of evolution, that we are in effect the purpose of it all.

Although we usually fail to think of it in this way, the world around us today is just one of countless possible worlds. The millions of species of plants, animals, and insects we see around us are the expression of myriad interacting processes, including chance—perhaps especially including chance. At any point in its prehistory, a species might just as easily have taken a different direction, given a slightly altered confluence of events, thus leaving today's world a slightly different place. And this includes the line leading to us. If, for instance, the massive asteroid collision that appears to have spelled the end of the dinosaurs had also wiped out completely the infant primate lineage that existed 65 million years ago, then there would have been no bush babies and other prosimians, no monkeys, no apes—and no us. And if the climatic changes that so altered the African landscape between 5 and 10 million years ago had in fact not occurred, apes might have remained the highest of the primate order, as they were then.

There are so many "ifs" in our history that could so easily have shifted the course of events. Despite our intense desire to believe otherwise, *Homo sapiens* simply cannot be seen as the inevitable product of life on earth.

But because accounts of human origins tell the beginnings of the story with half an eye to the ending, there is created, says Landau, "the myth of the turning point."[23] If, as seems the case with the benefit of hindsight, human origin is the exchange of a successful life in the trees for a successful life on the ground—an exchange Gregory terms "this momentous transformation"[24]—then indeed our ancestors did traverse a kind of rite of passage. There was a crucial transition, a real turning point, when a journey was being undertaken from one state (in the present) toward another (in the future). And each step along the way is just a part of a deliberate progression, one leading inexorably to the next. Or so it must seductively seem.

"Telling a story does not consist simply in adding episodes to one another," says Landau. "It consists in creating relations between events."[25] Consider the point at which our ancestors came to the ground. "Paleoanthropologists still do not agree about how, when and why it occurred. But to listen to them is to hear that however it happened, coming to the ground was a 'departure,' a 'decisive step' in human evolution." It is easy to see how, with a generous dose of anthropomorphizing, authors portray the event as a hazardous exercise: a defenseless ape faces the predatory hazards of the plains. This sentiment dominated paleoanthropological thought for many decades, beginning with Darwin, who sought to equip our ancestors with sharp stones and a sharper wit with which to defend themselves. But, argues Landau, "There is nothing inherently transitional about the descent to the ground, however momentous the occasion. . . . It only acquires such value in relation to our overall conception of the course of human evolution."

This notion of an animal committed to a journey with a clearly defined end is frequently expressed overtly in the earlier writings, as we've seen with Elliot Smith's words particularly. But it is also to be found with modern authors. Take, for instance, a recent interpretation of the way that Lucy and her fellows—*Australopithecus afarensis*—might have moved around. As part of one of the most interesting and lively modern debates in paleoanthropology, Jack Stern and Randall Sussman, of the State University of

New York at Stony Brook, wrote the following in the *American Journal of Physical Anthropology*: "In our opinion *A. afarensis* is very close to a 'missing link.' It possesses a combination of traits entirely appropriate for an animal that had traveled well down the road toward full-time bipedality, but which retained structural features that enabled it to use the trees efficiently for feeding, resting, sleeping or escape."[26]

"Metaphors cast powerful spells," comments Landau, "not only in everyday life but also in science . . . When Stern and Sussman say that '*A. afarensis* had traveled well down the road toward full-time bipedality,' not only do they speak in metaphor, they also tell a story."[27]

Lucy's anatomy, according to Stern and Sussman, was apparently adapted to a good deal of tree-climbing as well as walking on two feet while she was on the ground. She was clearly a different animal—quite apart from her diminutive brainpower—from modern humans in the way she moved around her world. Later, it seems, her evolutionary descendants were fully committed bipeds, like you and me. With this in mind, Stern and Sussman fall into the trap of describing Lucy's adaptations as transitional, as being on the way to the next stage in the story. The fact is that Lucy's mode of locomotion—a mixture of arboreality and terrestriality—was a perfectly good adaptation which might well have persisted for millions of years longer than it apparently did. It just didn't, as things turned out; that's all. There was nothing inevitable about the emergence of fully committed bipedalism in the evolution of large primates.

Stern and Sussman are by no means alone in the teleological trap, by the way: they have the company of virtually every paleoanthropologist who has put pen to paper on the subject of human origins.

Because of our natural and usually unrestrained vanity, the transformation of ape ancestor into *Homo sapiens* is viewed as the most splendid example of what is generally seen to be the epitome of evolution: progress. Like an arrow through time, evolution has been considered to be engaged in a constant process of improvement of form and function, clever adaptations constantly being honed to be yet cleverer. Keith puts it best in the paleoanthropological context: "In all these journeyings into ancient times and to primitive people there is one adage, an article of Darwinian faith, which we must bear in mind. Nature is jealous of her species building. Progress—or what is the same thing, Evolution—is her religion; the production of new species is her form of worship. She is up to every trick in this game she plays with living things."

The worship of progress in the world is in fact a feature peculiar to Western civilization, born initially of the tremendous material advances that accrued through the engines of the industrial revolution. Change; constant change and advancement—this became the operating ethos. And no doubt it was one reason the notion of Darwinian evolution became so acceptable to Victorian England: because it could be shaped within the same mold. "The myth of progress"[28] is how two scientists at the American Museum of Natural History characterize it. "Change is difficult and rare, rather than inevitable and continual," write Niles Eldredge and Ian Tattersall. "Once evolved," they comment, "species with their own peculiar adaptations, behaviors, and genetic systems are remarkably conservative, often remaining unchanged for several million years. In this light," they assert, "it is wrong to see evolution, or for that matter human history, as a constant progression, slow or otherwise."

If the idea of the inevitability of the process of producing humanity is a persistent theme of paleoanthropological writings, then so too is the notion that creation of *Homo sapiens* is the reason for the evolutionary process to be operating at all. This viewpoint, however, is much more that of Osborn and his contemporaries than of more recent practitioners. As always, Elliot Smith can be relied upon to express colorfully what others felt strongly. "The vast continents of Africa and Asia represented . . . the domain of Primitive Man during the early history of the Human Family and the laboratory in which, for untold ages, Nature was making her great experiments to achieve the transmutation of the base substance of some brutal Ape into the divine form of man."[29] In this one short statement is encompassed the notion of man's revered status in the world, the ape's inferiority, and Nature's goal in transforming the one into the other. Osborn clearly concurs in Elliot Smith's assessment of humanity: "As man is crowned king of the rest of the animal kingdom, the evolution of man is certainly a matter of supreme interest."[30]

Nature, however, was apparently not equipped with a clear-cut design for its ultimate goal: man. According to most writers, it had to engage in a good deal of trial and error before the intended divine form finally arose. "After many experimental types of the Human

Family had occupied the world for thousands of years, the genus *Homo* emerged," notes Elliot Smith.[31] "If the fossil finds . . . hitherto brought to light mean anything at all, they mean that Nature has conducted many and varied experiments upon the higher primates," writes Earnest Hooton. Keith's observation on the matter is the following: "We human beings are the subject of [Nature's] experiments—the pawns of her great game." These and other authors were writing at a time when the element of chance in Darwinian theory was in low repute, which explains in part the notion that the variety of prehuman types was the result of active experimentation rather than the result of random processes. This would have been the case whatever the organism under consideration. But as it was the origin of *Homo sapiens* that was at issue, few had any taste at all for accommodating the idea that their existence might have depended on capricious chance.

No one articulated the sentiment more directly, however, than Robert Broom, a Scotsman who helped pioneer the discovery of early human fossils in South Africa during the 1930s to the 1950s. "Surely there can be no subject so interesting to man as why he has appeared on earth,"[32] wrote Broom in 1933. "Much of evolution looks as if it had been planned to result in man, and in other animals and plants to make the world a suitable place for him to dwell in."[33] Once Nature had done its job, says Broom, the process stopped. "The evolutionary clock has so completely run down that it is very doubtful if a single new genus has appeared on earth in the last two million years,"[34] Broom observed inaccurately. Nevertheless, this led him to the conclusion that "there was no need for further evolution after man had appeared." Broom's position might be thought to be a little extreme, especially since he further concluded that "the evolution of man must have been deliberately planned by some spiritual power." But in fact, it was unusual only in the very forthright way in which it was stated. Osborn, Elliot Smith, and others were undoubtedly close spiritual fellow travelers with Broom.

During her reading of the paleoanthropological literature, Landau was intrigued not only by these grand narratives and philosophical assumptions, but also by the way in which practitioners talked about the hard evidence: the fossils. And here one can focus much more closely on current writings. There is a strong tendency, she claims, for fossils to be presented as if they were lucid texts to be read unambiguously, rather than scrappy fragments of unknown morphologies to be interpreted. "Let the fossils speak for themselves" is a phrase that's frequently spoken or written. Moreover, even when fossils are being described in the most technical terms, authors often invest their words with unspoken arguments: there are texts beneath the texts, she suggests. "The question to ask, then, is not what do fossils tell us about human evolution but what is it about human evolution—and not only human evolution—that through fossils is getting said."[35]

When Osborn denoted prehuman fossils as "these precious documents" he was articulating an attitude widely held in his time and often still expressed today. All that was required of the paleoanthropologists was that they read the documents correctly. "There are riches of knowledge and information locked up in these four fossils which could be released through the application of the most modern methods of research,"[36] Osborn said in 1921, referring to Piltdown and other material. It is as if the fossils were some kind of crude ore from which the ultimate truth could be extracted in pure form by the appropriate smelting process.

The concept that fossils literally speak for themselves is most graphically illustrated by the cover of Richard Leakey's best-selling book *Origins*, published in 1977. Superimposed on a sweeping view of the East African savanna is the face of the skull 1470, whose discovery in 1972 made Leakey famous. Beneath the skull are the words: "What new discoveries reveal about the emergence of our species and its possible future." Stirring stuff. "Admittedly, few anthropologists would claim to pay serious attention to blurbs or to book covers," says Landau.[37] "Still, I would argue that this picture represents reasonably, if garishly, what is implied if not stated in the most technical scientific discussions: that theories of human evolution are governed or, in a metaphorical sense, 'dictated' by fossils."

In fact, "virtually all our theories about human origins were relatively unconstrained by fossil data," observes David Pilbeam.[38] "The theories are . . . fossil-free or in some cases even fossil-proof."[39] This shocking statement simply means that there is and always has been far more fleshing out of the course and cause of human evolution than can fully be justified by the scrappy skeleton provided by the fossils. As a result, he continues, "our theories have often said far more about the theorists than they have about what actually happened."[40]

A good example of this is the emphatic shift in theoretical

stance between the 1950s and '60s, when the specter of Man the Hunter, Man the Killer Ape dominated paleoanthropology, and the 1970s and '80s, when peace and cooperation were stressed instead, with the emergence of Man the Social Animal. New fossils discovered during this transition made no contribution to the altered prevailing theory. But the social climate had taken a dramatic swing, from a time when war was an acceptable instrument of international policy to a time of realization that another such excursion on a worldwide scale could obliterate life on the planet. "When people turn indignantly from one sort of speculation to embrace another, there are usually good, nonscientific reasons for it,"[41] observes Duke University anthropologist Matt Cartmill. Paleoanthropologists were looking to their theories to explain the world as they saw it and the way they hoped it would be. Pilbeam once found a quotation that expresses this very well:

We do not see things the way they are;
We see them the way we are.

He thought it came from the Talmud, only later to discover that instead it was from a Chinese fortune cookie. Its source does not, however, diminish its force.

There are, however, according to Pilbeam, signs of improvement in paleoanthropology: "A major change is a growing realization that many evolutionary schemes are in fact dominated by theoretical assumptions that are largely divorced from data derived from fossils, and that many assumptions have remained implicit."[42]

Nevertheless, argument disguised—probably unconsciously—as objective description is still to be found in the literature, says Landau. She first looked at some earlier passages, including the announcement in 1925 by Raymond Dart of the first early human fossil to be found in Africa, *Australopithecus africanus*, or the Taung child, as it is often called. By convention, when a scientist reports the discovery of a new fossil, the paper includes a "description" that is meant to be a detailed picture in words, an objective projection of the physical characteristics of the fossil. When Dart reported the discovery of the Taung child in the English journal *Nature*, he wrote: "The orbits are *not in any sense* detached from the forehead, which *rises steadily* from their margins in a fashion *amazingly human*. The interorbital width is *very small* (13 mm.) and the ethmoids are *not blown out laterally as in modern African anthropoids.* . . . The molars, zygomatic arches, maxillae, and mandible all *betray a delicate humanoid character*"[43] (Landau's emphasis added).

It is clear from this short passage that Dart was doing more than piecing together a technical diagram in words. He was arguing that the Taung child was more human than ape, but doing it in the context of simple description. When Dart made his announcement of *Australopithecus*, the paleoanthropological community was wedded to a particular view of human origins against which the Taung child would make no sense at all: it was simply too primitive, too much like an ape. Perhaps Dart was aware that his proposal would be coolly received, and therefore tried to sharpen his point by overemphasizing the humanlike features of his truly ape-like fossil. Whatever was the case, for Landau the passage is a good example of a usually unmentioned aspect of paleoanthropological description: "namely, that it is thick with interpretation not about what the fossils look like but also about what they mean."[44]

When Don Johanson and his Berkeley colleague Tim White presented a major paper in the American journal *Science* in 1979 about the implications of Lucy and the other fossils from Ethiopia, they included a "description" of some of the salient features. Once again Landau highlights some elements in it: "The lower molars, particularly the first and second, *tend* to be square in outline. The cusps are *usually* arranged in a simple Y-shape pattern, surrounding wide occlusal foveae. The third molars are *generally* larger and their distal outlines are rounded"[45] (Landau's emphasis added). The immediate impression is of a more scientific approach than Dart's, and Johanson and White explicitly claim that they are separating description from theory. "As they make clear . . . through the passive constructions and detonalized voice which echo throughout their paper, Johanson and White are concerned as much with being objective as with naming fossils,"[46] Landau observes.

Nevertheless, she contends, it has to be realized that in publishing their paper Johanson and White were doing more than just describing their fossils and proposing a new family tree. They were making a major contribution to a stormy debate about what the Ethiopian fossils really are. One aspect of the debate was that although Johanson and White had suggested that all the fossils from their Hadar site in Ethiopia belonged to one hominid species, *Australopithecus afarensis*, some authorities were arguing that in fact there were two species represented by this large collection of fossils, and Johanson and White had made a mistake. So, running beneath Johanson and White's apparently objective description of Lucy's teeth in *Science*, says Landau, is an argument that "the molars belong to one species, *A. afarensis*, instead of two, *Australopithecus* and *Homo*, as some other paleoanthropologists believe." The argument goes beyond simply counting species, because it affects the shape of the family tree that can be drawn from their conclusions. In other words, Johanson and White couch their objective description in terms that stress their conclusions: namely, the unity of the fossils as a single species. Hence their generalizing words noted in their description of the fossil teeth. It is a largely unconscious process, says Landau—and everybody does it.

Landau's novel focus on the use of language in paleoanthropology—both on the grand scale of narrative form and in the nuance of fossil description—has undoubtedly made many researchers somewhat defensive. It seems to be a threat to the legitimacy of the science. But once again, this is in part because of the idealized image that scientists project of what they do: that elusive "objective search for the truth." Telling stories seems quite outside such a revered activity. But, insist Eldredge and Tattersall, "science *is* storytelling, albeit of a very special kind." And paleoanthropology is a science of a very special kind, too. This is partly because it is historical, and therefore is particularly susceptible to storytelling —but mostly because, in this materialistic world, it is meant to explain how we came to be here.

John Durant, a researcher at Oxford University, England, put it this way: "Like the Judaeo-Christian myths they so largely replaced, theories of human evolution are first and foremost stories about the appearance of man on earth and the institution of society."[47]

CHAPTER 3

THE TAUNG CHILD: REJECTION

In February 1985, Professor Raymond A. Dart celebrated both his ninety-second birthday and the diamond jubilee of his announcement to the world of the Taung child, a diminutive fossilized skull whose discovery revolutionized man's search for his origins. Two hundred of the world's most prominent anthropologists joined Dart in his hometown, Johannesburg, South Africa, to mark the event and to pay homage to the man whose insight sixty years earlier opened a new era in paleoanthropology. It was an era in which anthropologists came to accept Africa as the cradle of mankind, just as Charles Darwin had predicted more than a century ago.

Ten days of scientific symposia marked the Taung diamond jubilee, and they put on display the great advances of the past six decades. Included was the announcement that the Taung child had in fact been three years old when he died, not six years as had previously been believed.

"What a wonderful occasion this is, isn't it?" says Dart. Then, after a moment's reflection, he continues: "You know, I was never bitter about how I was treated back in 1925. I knew people wouldn't believe me. I wasn't in a hurry."[1] He laughs, and you know that a less resourceful, less independent man would have been spiritually broken by what Dart had endured. Serious again, he adds, "I just wish that man Zuckerman could be here to see all this." At which point he turns and, aided by his wife's arm, walks slowly away. And suddenly you know how difficult it really was, sixty years ago.

Sherwood Washburn, who recently retired as professor of anthropology at the University of California, Berkeley, makes the following observation: "In the 1920s there were very few human fossils. Scientists were eager to discover more, and one might therefore think that the discovery of a new kind of human being would have been greeted with joy and interest. But in fact, the situation was exactly the reverse. Most of the leading scientists of the day criticized Dart, both for his description of the fossil and for his evolutionary conclusions."[2]

The Taung child is a jewel of a fossil. Its face is intact, and carries the upward sweep of the cranial vault. A complete lower jaw shows its young teeth already well developed. And perhaps most striking of all, the fossil included a natural endocast, a petrified replica of the shape of the child's brain as it impressed itself on the inner surface of the cranium, its convolutions and contours standing out clearly even to the untrained eye. It is a rare document from the usually unyielding fossil record. Cradle the little fossil in the palm of your hand, look into its now-blank eye sockets, and you get a feeling of staring way back into your own past, a feeling that happens only rarely with other prehuman fossils. There is something special about the Taung child, and the young Raymond Dart, thirty-two at the time of the discovery, immediately recognized it, both in the strict realms of science and in the emotional dimension too.

What Dart saw was that although the fossil had the face of an ape, its brain was that of a human—not in size, but in some key elements of its architecture. Had the Taung child grown to adulthood, its brain would have expanded to about 450 cubic centimeters in capacity, which is the size of a gorilla's and not much more than a third that of a modern human's. Nevertheless, Dart was an expert in neurology, and he believed he could recognize a human brain—even an incipient human brain—when he saw one. "That's what put me on to the idea that the fossil wasn't just an ape's," he now says. "Without that endocast, and without my experience in neurology, I doubt I would have thought it was a hominid."[3] But alerted by traces of what he took to be humanlike contours in the endocast, Dart went on to notice that the head had been balanced atop the vertebral column, as in humans, and not slung forward, as in apes; the clue here is the position of the foramen magnum, which is the aperture through which the spinal cord leaves the cranium and enters the spinal column. In other words, the Taung child had been bipedal: it had walked on two legs, not four. Here, then, was a creature that had a brain the size of an ape's but with hints of emergent humanity in its architecture; had an ape's face, and yet walked on two legs, like a man.

Having identified the anthropological uniqueness of the Taung child, Dart worked intensively to clean and prepare the fossil for scientific description and publication. In his now classic 1925 paper he said that the Taung child was a member of "an extinct race of apes intermediate between living anthropoids and man," and so proposed a new family—Homo-simiadae—into which to place it. This new family, which never caught on in scientific circles, was meant to fill the previously unoccupied gap between humans and apes. Nothing as primitive as the Taung child had ever been proposed as belonging to the human family. The formal genus and species name Dart chose for the fossil was *Australopithecus africanus,* or the southern ape from Africa.

Dart had first seen the fossil on Friday 28 November 1924, when it was brought to his Johannesburg home from the Taung lime quarry in the southwestern corner of the Transvaal, near the diamond town of Kimberley. Just forty days later, on 6 January 1925, he mailed his bombshell of a manuscript to the British journal *Nature,* which is rapid progress by any standards. Too rapid, as it turned out, by about twenty-two years.

Dart had not gone to South Africa with the intention of finding the "missing link." He was not driven by an urge to find his ancestors and a conviction of where they might be—unlike the Dutchman Eugène Dubois, who had gone to Java thirty years earlier with just such a mission in mind. In fact, Dart had gone to work in South Africa with great reluctance.

Australian by birth, Dart had spent two intellectually stimulating years in London under the tutorship of the eminent British neuroanatomist Sir Grafton Elliot Smith. Socially and professionally intimate with that other notable name of early twentieth-century British anatomy Sir Arthur Keith, Dart was thrilled and excited by the rare scholarship with which he found himself so closely associated. Unorthodox and a maverick by nature, Dart nevertheless wanted nothing more than to be in London doing science with these great men. Dart's first love was neuroanatomy, but like Darwin, he was a man of enlarged curiosity, and he greedily consumed the wide range of intellectual fare that was on offer in this scholarly group. In addition to their stature as anatomists, Elliot Smith and Keith were, of course, key figures in the British anthropological establishment.

Given all this, it is not difficult to imagine Dart's reaction when, in mid 1922, Elliot Smith encouraged him to apply for the newly created chair of anatomy at the University of the Witwatersrand Medical School in Johannesburg. "I thought it was an appalling idea," Dart recalls. "I thought South Africa was a horrible place to be, and it would be so isolated from the scholarship of London."[4] But both Elliot Smith and Keith were very persuasive, arguing that such a move would be good for Dart's career. So, just before Christmas of 1922, Dart sailed for South Africa, with little enthusiasm in his breast and the daunting prospect ahead of him of establishing more or less single-handedly and with meager resources the anatomy department of the University of the Witwatersrand Medical School in Johannesburg. Keith was later to write of Dart's move: "I was the one who recommended him for the post, but I did so, I am now free to confess, with a certain degree of trepidation. Of his knowledge, his power of intellect, and of imagination there could be no question; what rather frightened me was his flightiness, his scorn for accepted opinion, the unorthodoxy of his outlook."[5]

So it was that just two years after Dart arrived in South Africa to pursue what immediately proved to be an active and extremely productive career as a neuroanatomist, he would be dispatching a manuscript on anthropology back to his beloved London, and, indirectly, into the hands of his former mentors Elliot Smith and Keith. When the editor of *Nature* received Dart's Taung manuscript on 30 January 1925 he immediately realized it would provoke wide discussion and so, just four days later, sent proof copies to Elliot Smith, Keith, and two other leading British anthropologists whose comments he wished to publish. Dart's article, accompanied by an absurdly tiny picture of the skull, appeared in the issue of 7 February, and the Taung child was instantly hailed in the popular press as a "missing link"—an appellation of which the newspapers have apparently always been fond. It wouldn't be long, however, before the Taung skull became the butt of cartoons and music-hall jokes.

Serious commentary, however, began in the following issue of *Nature*, dated 14 February, in which Elliot Smith, Keith, Sir Arthur Smith Woodward, and Dr. W. L. H. Duckworth had their say. Dart had not been expecting instant acceptance of his interpretation of the Taung fossil, but he was rather disappointed that the reaction was so very negative. "It may be," opined Keith, "that *Australopithecus* does turn out to be 'intermediate between living anthropoids and man,' but on the evidence now produced one is inclined to place *Australopithecus* in the same group or sub-family as the chimpanzee and gorilla." Elliot Smith was equivocal: "Until Prof. Dart provides us with fuller information and full-sized photographs revealing the details of the object, one is not justified in drawing any final conclusions as to the significance of the evidence." Smith Woodward was the most critical: "It is premature to express any opinion as to whether the direct ancestors of man are to be sought in Asia or Africa. The new fossil from Africa certainly has little bearing on the question." While not strongly supportive of Dart, Duckworth was by far the most positive of the four commentators. He was, it is worth noting, also the only one not to be knighted by the establishment.

Rather than mellowing with the passage of time, these gentlemen's remarks became more and more hostile to Dart and his fossil. For instance, in a lecture at University College, London, four months later, Elliot Smith said the following: "It is unfortunate that Dart had no access to skulls of infant chimpanzees, gorillas or orangs of an age corresponding to that of the Taung skull, for had such material been available he would have realized that the posture and poise of the head, the shape of the jaws, and many details of the nose, face, and cranium upon which he relied for proof of his contention that *Australopithecus* was nearly akin to man, were essentially identical with the conditions in the infant gorilla and chimpanzee." The Taung skull, Elliot Smith was making plain, was simply that of a young ape.

When Arthur Keith, later that summer, first saw a plaster cast of the skull, he said in a statement to the press, "The famous Taung skull is not that of the missing link between ape and man." In a letter to *Nature* penned on 22 June he wrote the following comment on Dart's suggestion that the Taung child was midway between ape and human: "an examination of the casts . . . will satisfy zoologists that this claim is preposterous. The skull is that of a young anthropoid ape—one which was in its fourth year of growth, a child—and showing so many points of affinity with the two living African anthropoids, the gorilla and chimpanzee, that there cannot be a moment's hesitation in placing the fossil form in this living group." Keith also charged that Dart's interpretation of the human features in the endocast was "a matter of guess-work." Elliot Smith and Keith were evidently in solid agreement that their former student had made a sorry mistake. Keith's trepidation at recommending Dart for the Johannesburg professorship had, it seemed, been justified.

Keith's negative interpretation was clearly a further development of his earlier remarks, but its trenchancy probably had not a little to do with the fact that he had been forced to view the cast through a glass case along with ordinary members of the general public while it was on display at the British Empire Exhibition at Wembley, just outside London. "For some reason, which has not been made clear, students of fossil men have not been given an opportunity of purchasing these casts; if they wish to study them they must visit Wembley and peer at them in a glass case," he snorted. "Yes, I know that Keith was very cross about that,"[6] Dart now recalls.

So, the British establishment firmly set its face against Dart's interpretation of the Taung child, a fossil that is now recognized by the great majority of leading anthropologists as having played an important role in human prehistory. And the same climate prevailed in the United States. "Perhaps the most widely read book of the time was Hooton's Up from the Ape (1931)," notes Sherwood Washburn. "But there was nothing about Australopithecus, Taung or Dart."[7] William Howells, who recently retired as professor of anthropology at Harvard University, remembers that "As undergraduates in those days, we heard nothing about Australopithecus in our courses. No scuttlebutt among enterprising or curious graduate students. Nothing."[8] Why? Why, when Elliot Smith and his contemporaries examined the Taung child's anatomy, did they read "ape," with no shades of humanity whatsoever?

One problem that apparently hindered the acceptance of the Taung child as relevant to human ancestry was that it was simply in the wrong part of the world. Although Charles Darwin had stated in his Descent of Man, which was published in 1871, that Africa was the most likely continent on which to find early human ancestors, the idea was distinctly out of fashion in 1925. Instead,

Asia was considered by most to be the center of it all, a sentiment clearly expressed by Richard Swann Lull, professor of anthropology at Yale University.

"That Asia is the birthplace of mankind is seemingly established," wrote Lull in 1921. "Asia possesses great size, and hence varying life conditions, together with a central location contiguous to all other land masses, even, as the north polar projection shows, to North America. . . . Asia is the home of the highest and best of organic life and is with few exceptions the place where man has derived his dependents and allies, domestic animals and plants. Asia is the seat of the oldest civilizations, many indications of which, though visible as sand-drifted ruins, have outlived the vaguest traditions concerning their origins. Finally, physical and climatic conditions of Asia in the Tertiary era were such as the scientist must postulate in his imaginings of the *modus operandi* of human origin from his prehuman forbears, i.e., such as would enforce descent from the trees and terrestrial adaptation."[9]

This fixation on Asia seduced Henry Fairfield Osborn, aristocratic director of the American Museum of Natural History, New York, into launching several spectacular and ambitious expeditions to the Gobi Desert in search of the first men; his intrepid explorers returned with dinosaur eggs, but no early men. And it also led to the very ready acceptance into the human family of the most meager of fossil evidence—a single tooth—which first came out of Chou Kou Tien, the famous Peking Man site, in 1926. "The immediate reaction to the discoveries in China was appreciation, encouragement and generous financial support," comments Washburn. "Far from the bones being objective facts to be judged as evidence, there was an established pattern of belief. There was a climate of opinion that favored discoveries made in Asia but not the 'silly notion' of small-brained bipeds from Africa."[10]

So geography was one problem for Dart; but it was only one of many. For instance, Darwin had based his African prediction on what he saw as a close evolutionary relationship between humans and the African apes. "In each great region of the world, the living mammals are closely related to the evolved species of the same region," Darwin had written in 1871. "It is, therefore, probable that Africa was formerly inhabited by extinct apes closely allied to the gorilla and chimpanzee: and as these two species are now man's nearest allies, it is somewhat more probable that our early progenitors lived on the African continent than elsewhere." To find a distinctly apelike human ancestor in Africa, such as Dart claimed he had done, is just what Darwin would have expected.

Darwin's argument on geography and descent is simple, logical and persuasive—that is, unless it happens to offend one's sensibilities. And in the 1920s it did just that for many professional anthropologists, not to mention the less scientific members of the public. Nineteen twenty-five, the year of Taung, remember, was also the year of the Scopes trial in Dayton, Tennessee. Evolution was not a popular subject in the United States, still less an admission of close familial relationship with a chimpanzee. And for many people, Dart's Taung child was just too much like a chimpanzee to be admitted into the human pedigree. Clearly, the best way to exclude an apelike creature even from consideration as belonging to the human family was to see it as all ape, with nothing humanlike about it at all. Such a judgment is, of course, subconscious, but, being deeply rooted in fertile emotional soil, is nevertheless strongly expressed.

The idea of finding an ape in our family tree was particularly stridently debated in the United States, where Henry Fairfield Osborn and his American Museum of Natural History colleague William King Gregory openly disagreed on the matter for years.

In his position as director of the American Museum, Osborn was the natural spokesman in favor of evolution against the likes of William Jennings Bryan, who was the chief prosecutor at the Scopes trial. This Osborn carried out with enormous exuberance, bringing his powerful personality to bear on every public medium he could exploit. He wrote articles for *The New York Times* and made frequent radio broadcasts, berating the fundamentalists' attacks on evolutionary theory. And he was delighted to be able to claim evidence for the existence of early human progenitors in the United States—namely, a tooth discovered by paleontologist Harold Cook early in 1922 in Bryan's own home state, Nebraska. Osborn, being an orator of some considerable force, if not the subtle talent of Bryan, was able to make great play of that association. Alluding to the quotation from the Book of Job (12:8) "Speak to the earth, and it shall teach thee," Osborn wrote in 1925: "The earth spoke to Bryan, and spoke from his own native state of Nebraska."

Named by Osborn *Hesperopithecus haroldcookii,* or more popularly Nebraska Man, the tooth was considered to be that of a very early anthropoid which had inhabited the shadowy beginnings of humanity. For Osborn, a devout man, its discovery was an answer to his prayers in the cause of evolution. Eventually, however, the tooth was shown to be that of a peccary (a kind of pig), and not that of an anthropoid at all, which revelation proved to be more than a little embarrassing for the museum's director.

"In this atmosphere, when Osborn could use a single, badly worn tooth to bedevil the anti-evolutionists with the improbable notion of a Pliocene ape-man in America," comments William Howells, "it is difficult to see why Dart's explicit, detailed and highly suggestive paper was not avidly seized on as a better club to bash the bumpkins. But it was not."[11]

But although Osborn was the chief public defender of evolution in the United States, it was not Darwin's evolution he was defending: it was a very aristocratic view of the world, and of humans in particular. The whole system, he said, was driven by effort, whose reward was progress and in the end a clear superiority for a few. As a result, an immense gulf separated humankind from the rest of the animal kingdom, and no small gulf divided the "superior" from the "inferior" races of humanity. Naturally, there was absolutely no question about which rung of the ladder he, Osborn, occupied. Racism of a peculiarly pure, intellectual form was a persistent theme of American and British anthropology of the time, and not surprisingly Osborn was a leading figure in the eugenics movement. Consequently Osborn, and to a lesser extent Keith, had a very arrogant view of the world, in which the evolution of man was a noble undertaking, certainly with no place for close relations with a tree-climbing ape. Keith later shifted his views, but Osborn never did.

Darwin's view of human origins, which was supported by Thomas Henry Huxley in his own country and Ernst Haeckel in Germany, had remained part of the currency of evolutionary theory through the end of the nineteenth century, but began to fade as, during the first four decades of the twentieth century, the pure form of Darwinism itself fell temporarily out of favor with the scientific establishment. The shift of focus from Africa to Asia as the center of human origins was part of that intellectual movement. And for Osborn, the combination of Asia and the necessary nobility of human origins proved to be an intoxicating cocktail.

In October 1923, when Osborn joined one of the American Museum's expeditions to Mongolia, he had a revelation whose grand scale and swaggering style say more than words could about their author: "I suddenly found myself forming an entirely new concept of human origins, namely that the actual as well as ideal environment of the ancestors was not in warm forested lowlands . . . but in the relatively high, invigorating uplands of a country such as Asia was in the Miocene and Oligocene time—a country totally unfitted for any form of anthropoid ape, a country of meandering streams, sparse forests, intervening plains and meadowlands. Here alone are rapidly moving quadrupedal and bipedal types evolved; here alone is there a premium of rapid observation, on alert and skillful avoidance of enemies; here alone could the ancestors of man find materials and early acquire the art of fashioning flint and other tools." [12]

On returning to Peking, Osborn, in an exhilarated mood, delivered to the association of Wen Yu Hui (Friends of Literature) an extemporaneous address entitled "Why Mongolia May Be the Home of Primitive Man." He proclaimed: "We observe that early man was not a forest-living animal, for in forested lands the evolution of man is exceedingly slow, in fact there is retrogression, as plentifully evidenced in forest-living races of today. Those South American Indians who live in forests are backward in development compared with those living in the open. Of the latter, those living in the uplands are more advanced than those living in the river drifts." [13] Here, then, is a cameo of Osborn's world that is comfortingly consonant with his imperialist view of some of its "inferior" races.

Osborn gave an account of all this in his book *Man Rises to Parnassus*—a title that again reveals its author's sense of human origins. Completed in February 1927, exactly two years after the publication of the Taung skull in *Nature*, the book contains not a single reference to Dart, Taung, or *Australopithecus africanus*. It wasn't that Osborn dismissed the idea of anthropoids' having anything at all to do with human history. Rather, he simply refused to accept that the human line had ever passed through a stage that might in any way be equated with modern chimpanzees and gorillas, as Darwin said and Osborn's friend and colleague Gregory seemed to imply. As long as the anthropoid in our past could be kept at a distance—and the greater the better—Osborn was happy. In fact, in July 1927 he wrote to Arthur Keith, "I am perfectly confident that when our Oligocene ancestor is found, it will not be an ape, but it will be surprisingly pro-human." Although the geological periods were still somewhat uncertainly dated in Osborn and Keith's time, the Oligocene was about as far back as one could go (some 30 million years in today's terms) and still expect to find a true human ancestor of any kind.

The result of all this for Osborn was that "We are forced to reconsider Darwin's concept of the primitive ape-man as inhabiting a 'warm, forest-clad land,' "[14] he told a gathering of the American Association for the Advancement of Science in December 1929. Explaining why he thought the origin of true man should be sought way back in the geologic record, he said, "To my mind the human brain is the most marvelous and mysterious object in the whole universe and no geologic period seems too long to allow for its natural evolution."[15] Sir Arthur Keith, incidentally, once expressed a similar sentiment. As a replacement for Darwin's Ape-Man, Osborn offered instead what he called the Dawn-Man, a creature who, though less than modern man, was nevertheless not far distant from him. One of the enigmatic aspects of Osborn's image of his ancestor is that even as one passes back through eons of time, his Dawn-Man remains surprisingly modern in form. What more primitive form existed prior to the Dawn-Man is never made entirely clear. It is as though Osborn couldn't bring himself to acknowledge a primitive, apelike form as an immediate ancestor of early true man.

Osborn was evidently greatly comforted by what his Dawn-Man theory appeared to offer, and he thought others would be too. "We have all borne with the monkey and ape hypothesis long enough and are glad to welcome this new idea of the aristocracy of man back to an even more remote period than the beginning of the Stone Age," he told a popular radio audience in February 1930. On another occasion Osborn proclaimed, "The most welcome gift from anthropology to humanity will be the banishment of the myth and bogie of ape-man ancestry and the substitution of a long line of ancestors of our own at the dividing point which separates the terrestrial from the arboreal line of primates."[16]

Not surprisingly, Osborn found a large and enthusiastic audience for statements such as these, and his correspondence files are full of letters praising his insight. "I send you my congratulations for being the first scientist of fame, who has the courage and Brain Abilities to question the theorys [sic] of Darwin-Spencer, Huxley," wrote a gentleman from California in December 1929. Another correspondent, this time from Pennsylvania, proclaimed that Osborn's new theory ". . . forever eradicates the possibility of associating man as being a process of evolution from the lower animals who inhabited forests and lived in trees." Osborn was frustrated at being "misunderstood," and yet he apparently failed to understand the illogicality of his own position.

For William King Gregory, Osborn's close colleague at the museum, the whole business was even more frustrating, for he was the major proponent at the time of the increasingly unpopular Ape-Man theory. Gregory's intellectual position is well stated in a postscript to a letter to Osborn, written 30 November 1920: "I fear we have reached opposite conclusions. . . . 'Back to Huxley and Darwin' is the motto of my conclusions." Yet in spite of their very deep differences of opinion, the two men were able to work closely together on museum business and conducted a very public but very polite debate over a period of many years. For instance, at the end of a public discussion at the museum in March 1927, Gregory said, "Professor Osborn has from the first shown the greatest generosity toward Dr. Hellman and myself in our differences of opinion."[17]

The two men wrote many popular and technical articles on the subject of the Ape-Man versus the Dawn-Man. They were jointly interviewed in magazines and on radio, and engaged many a lecture audience with their separate and equally strongly held convictions. As Gregory explained on one such occasion, "The great and critical difference between Professor Osborn's view and mine is the question of what relationships between man and apes really are and whether or not man passed through an arboreal, proanthropoid brachiating stage."[18] Gregory, an expert anatomist and well experienced in the details of the primate skeleton, said he could see many similarities between human and African great-ape anatomy that, to him, spoke of close evolutionary relationship, a recent common ancestry. Osborn, an expert anatomist on many groups of vertebrates but not primates, said that what similarities there were must be the result of parallel evolution, not common descent.

"It is a matter of sincere regret to me that I cannot follow my honored leader," said Gregory in a lecture before the Medical Society of the County of Kings early in 1927. "I deem it rather my duty to defend the old and always unpopular views of Darwin, Huxley and Haeckel. . . . I must attack his whole argument, in so far as it is based upon his studies of mammals other than primates, as a series of analogies, unsupported by direct evidence and outweighed by many definite facts of the record."[19] He contended that the Dawn-Man had simply been conjured up to satisfy an unfounded fantasy whose real source was a phobia. "It may be called pithecophobia, or the dread of apes—especially the dread of apes as relatives or ancestors," quipped Gregory.

So here we have two great scientists of their time, major figures in American anthropology: they looked at the same evidence and yet saw different things, primarily because one was using the lens of Huxley and Darwin while the other was gazing up to heights of Parnassus.

Although few American anthropologists expressed themselves as forcefully and in quite the same florid terms as Osborn, most were inclined more toward his view than toward Gregory's. The Taung baby could therefore not expect to receive an enthusiastic welcome from this group of professionals.

Meanwhile, the British anthropologists were heaping more and more contumely on Dart and his hypotheses. The date of the fossil, they said, was unknown and therefore its interpretation was impossible. Or it was too young in geological terms to fit as a protohuman into anyone's evolutionary scheme. The fossil was that of a child, and therefore it was dangerous to draw any kind of anatomical conclusions, ran another line of argument. In any case, Dart had not made the proper comparisons with apes of various stages of maturity. The language in his *Nature* paper, they sniffed, had been too florid for proper scientific discourse. And he had certainly been too hasty in reaching any kind of conclusion, certainly one of the magnitude that he did.

They even criticized his choice of name, *Australopithecus africanus*, combining as it does Greek and Latin roots. "It is generally felt that the name *Australopithecus* is an unpleasing hybrid as well as being etymologically incorrect," intoned an unsigned editorial in the 28 March 1925 issue of *Nature*. The editorial went on to lament that, to judge from Dart's reaction to this criticism, "the niceties of etymology do not generally appeal to him." On the other hand, according to University of Illinois anthropologist Charles Reed, "the International Rules of Nomenclature require only that a proposed new generic name consist of a unique combination of letters, so etymology is not involved."[20]

"It makes one rub one's eyes," wrote Robert Broom, a remarkable Scottish doctor and paleontologist who also worked in South Africa and for a long time was one of Dart's very few supporters.

"Here was a man who had made one of the greatest discoveries in the world's history—a discovery that may yet rank in importance with Darwin's 'Origin of Species'; and English culture treats him as if he had been a naughty schoolboy. . . . I was never able to discover what were Prof. Dart's offences: Presumably the most serious offence was that when he found a very important skull he did not immediately send it off to the British Museum, where it would have been examined by an 'expert,' and probably described ten years later, but boldly described it himself, and published an account within a few weeks of the discovery."[21]

Certainly, it is clear from the volume and diversity of the clamor from the halls of the British anthropological establishment that what Dart had said and the manner in which he had said it was all distinctly improper. But one of the great ironies of the rejection of the Taung child, of course, is that it was at Elliot Smith and Keith's instigation that Dart had gone to South Africa in the first place. Dart's literary style was indeed florid for scientific discourse, but then so was that of his mentor, Elliot Smith. And it was with the tools of the neurological trade bestowed upon him by Elliot Smith that Dart was able to palpate incipient humanity in the contours of the Taung baby's brain. But irony is heaped upon irony in this story, because it was this same focus of intellectual attention—Elliot Smith's expertise in human neurology—that helped put into place the major tangible obstacle blocking the Taung child's route to the human family tree. That obstacle was the infamous Piltdown Man, the "fossil" that held the British anthropological establishment in its thrall for nigh on four decades.

The various specimens that constitute the Piltdown Man were found over a period of half a dozen years, from 1912 onward, from two sites near the town of Piltdown, Sussex, just 40 kilometers from where Charles Darwin spent the greater part of his life. Their initial discoverer was Charles Dawson, a lawyer and amateur prehistorian. From the first site, a gravel pit, came several fragments of a remarkably humanlike cranium together with part of a remarkably apelike lower jaw. It was an extraordinary combination, and clearly begged the question of whether cranium and jaw had belonged to the same individual. The key part of the jaw joint that would have settled this question had been broken off and lost during the fossilization process, an unfortunate quirk of fate—or so it had seemed. Other fossils found with Piltdown appeared to indicate considerable antiquity, perhaps as early as the beginning of the Pleistocene (which would now be dated at 2 million years ago).

In the face of varying degrees of skepticism from North America and continental Europe, the British anthropological establishment concluded in near unanimity that jaw and cranium were indeed from one individual, that it was an ancient form of humanity, and what is more, its unusual form was precisely what would have been predicted, given prevailing theory. Virtually every major voice in British anthropology proclaimed that although the cranium was clearly very modern in aspect, many apelike features could also be discerned; and that while the jaw certainly looked like that of an ape, the trained eye could readily discern important human characteristics in it.

In fact, as was discovered a long forty years after its first announcement, the Piltdown Man was a hoax, a fraudulent seeding of the Piltdown gravel pits with fragments from a modern human cranium and an orangutan's jaw. Unsolved to this day, the Piltdown forgery remains one of the great whodunits of modern times.

The puzzle of the culprit's—or culprits'—identity has of course fascinated amateur historical sleuths for years, with the result that virtually everyone involved in the discovery and study of Piltdown has at some time or other been fingered as the perpetrator. As a result, says Michael Hammond, a sociologist of science at the University of Toronto, the real story of it all has been somewhat obscured: "namely, what could have led so many eminent scientists to embrace such a forgery?"[22] How is it that trained men, the greatest experts of their day, could look at a set of modern human bones—the cranial fragments—and "see" a clear simian signature in them; and "see" in an ape's jaw the unmistakable signs of humanity? The answers, inevitably, have to do with the scientists' expectations and their effects on the interpretation of data.

At the time that the Piltdown finds were announced, anthropology had been going through a set of theoretical changes into which the Piltdown fossils fitted as snugly as if they had been tailored that way—which, of course, they had. According to Hammond, these changes include the following: Arthur Keith's ideas on the great antiquity of man; Grafton Elliot Smith's hypotheses on the part played by brain expansion in human evolution; William Sollas' work on mosaic evolution, the idea that different parts of an organism might evolve at different rates; and Marcellin Boule's recent analysis of Neanderthal Man, in which he said the species had become extinct without descendants. "Due to their devotion to these new ideas," says Hammond, "a protective screen emerged around the forgery and played a crucial role in its initial acceptance and later defence."

Perhaps most fundamental of all these was the revolution in thinking about the Neanderthals, which Boule, an eminent French paleontologist, wrought almost single-handedly between 1908 and 1912. That the effect of Boule's conclusions was crucial to the eager acceptance of Piltdown as genuine, there is no doubt. That Boule's interpretations of the Neanderthal fossils he was studying were entirely erroneous, powered as they were by a particular set of preconceptions of his own, simply adds yet another layer of irony to our main story: namely, why the Taung child was dealt with so harshly when it was first offered as a member of the human family. To understand fully the rejection of Taung, you have to understand the acceptance of Piltdown, and in turn you have to understand the expulsion of the Neanderthals.

CHAPTER 4

THE TAUNG CHILD: ACCEPTANCE

The fossil bones that gave Neanderthal Man his name were recovered in 1856 from a cave being quarried for lime high up in a deep, narrow ravine known as the Neander Valley through which the Düssel River flows, a short distance from where it meets the Rhine at Düsseldorf, Germany. Overall, the fossils had a distinctive appearance about them: the skull bones were exceptionally thick, the browridges were unusually prominent, the limbs extremely robust, and the leg bones bowed. At the very least, Neanderthal Men were stockily built, powerful individuals.

The bones came into the hands of Hermann Schaffhausen, an anatomy professor, who considered that here was an individual who had belonged to an apparently barbarous population from the early pages of human history, perhaps one of the most ancient races of men. He first presented his ideas at a meeting of the Lower Rhine Medical and Natural History Society in Bonn on 4 February 1857, which was less than three years before Darwin published his *Origin of Species*.

The immediate reaction to the Neanderthals was mixed, not least because it was impossible to say from exactly what part of the geological past the bones came. Problems associated with incorrect or uncertain dating of fossils have been a constant irritant in paleoanthropology, and here was the first major example.

Mostly, however, people were responding to the apparently "uncouth" and "brutish" appearance of the Neanderthals. One German anatomist, rejecting the suggestion that the bones were ancient, said that the individual was a Mongolian cossack of the Russian cavalry which had pursued Napoleon back across the Rhine in 1814. The cossack, according to this expert, had become separated from his fellows, perhaps through injury, and had crawled into the cave to die. His bowed legs were clearly the result of a life on horseback. Another scholar interpreted the Neanderthal's bowed legs as the result of rickets, the pain from which disease had made the individual habitually furrow his brow, thus producing the prominent browridges. Still other scholars suggested that the Neanderthals might be forerunners of modern man, or at least related to the supposedly "inferior" races, such as the Australian aborigines.

When Rudolph Virchow, Germany's most eminent anatomist and pathologist of the time, proclaimed the Neanderthal to be of recent date, his unusual appearance being the result of pathology, including rickets, the argument seemed to be settled. Such is the weight of authority in any science, but particularly so in paleoanthropology, a science that is often short on data and long on opinion.

Eventually, however, Virchow's influence waned as more and more fossil bones were discovered, each of which was like those from the Neander Valley. Resort to pathology as an explanation of their appearance was clearly not tenable. By the turn of the century, when evolutionary theory finally became respectable in academic circles, the Neanderthal people were seen by some as having been a rung on the evolutionary ladder to modern man, and a rather recent one at that. The view of human evolution at the time was rather simple, with our ancestors being seen as occupying different stages on a straight-line progression from the primitive to the advanced: "passing step by step from a Simian to a modern type of man," as Keith put it at the time. It was perceived as a long, slow process, with Neanderthal being one of the last stages of premodern man and *Pithecanthropus erectus,* a more primitive form discovered by Eugène Dubois in Java in the early 1890s, one of the earliest of those stages.

But that old dating bogey raised its head again with the discovery of several specimens of clearly modern appearance but apparently ancient origin. One such was Galley Hill Man, a skeleton from England that Arthur Keith for quite some time believed was at least as old as and probably older than Neanderthal. Another was Grimaldi Man, from southern France, which Marcellin Boule considered to be of relatively modern form and yet virtually contemporary with Neanderthal. If these modern forms lived side by side with more primitive types, then the idea of a steady, unilinear progression must be false, Keith and Boule would argue.

By the end of the second decade of the twentieth century, the old image of a ladder of evolution came to be replaced by a bush, with many side branches that represented dead ends, extinct forms of failed, proto-men. Neanderthal came to be thought of as one of these dead ends, as did *Pithecanthropus* and indeed Piltdown. In fact, more or less every new humanlike fossil was shunted off onto a side branch, so that instead of every fossil being on the line to modern man, as previously envisaged, virtually none was. In effect, *Homo sapiens* came to be "ancestorless man," as one observer noted. Franz Weidenreich was intrigued by these developments and in 1943 made the provocative observation that what in fact one was seeing was "the last bastion from which the final acceptance of Darwin's theory could be warded off with a certain air of scientism."[1]

But in 1908, the concern about the apparent contemporaneity of modern forms with Neanderthal Man was just that: an unproved concern. "What was lacking was a well dated and relatively complete single skeleton that could settle the questions about the exact nature of Neanderthal morphology," explains Michael Hammond. "It was just such a skeleton that was unearthed in August 1908 by three clerical historians, the Abbés J. and F. Bouyssonie and L. Bardon. The skeleton was sent to Boule and arrived at the Museum of Natural History [in Paris] in early November 1908. La Chapelle-aux-Saints was the most complete Neanderthal skeleton discovered until that time."[2]

The skeleton might equally properly have gone to the École d'Anthropologie in Paris, except for that institution's anticlerical tradition. The Abbés Bouyssonie and Bardon were therefore happy to accept the Abbé Breuil's suggestion that instead it should be placed in the care of his friend Marcellin Boule at the museum. Had the Chapelle-aux-Saints skeleton been found by nonclerical discoverers, it might well have been sent to the École, and subsequent anthropological history might have taken a different course. As it was, Boule began a single-minded and intensive study of the skeleton, delivering his first presentation to the Academy of Sciences on 14 December 1908. This was followed by two more presentations, in May and June 1909, and a series of major publications through to 1912. The upshot of it all was that Boule clearly and unequivocally expelled the Neanderthals from human ancestry.

"Boule . . . depicted the Neanderthals in terms which have served journalists and scholars ever since as the basis for the caricature of the cave man," says University of Michigan anthropologist Loring Brace. "Since he was not prepared to accept such a creature in the human family tree, he settled the question to the general satisfaction by declaring that the Neanderthals . . . became extinct without issue."[3] In other words, people were more than ready to accept the suggestion that the supposedly "brutish" Neanderthals were not on the ancestral line to modern man.

Boule's motives for expelling Neanderthal from human ancestry were several and complex. For instance, it perhaps should be no great surprise that Boule would favor a bush over a ladder for the configuration of human evolution. After all, such had been his interpretation of the evolutionary pattern of many vertebrate groups he studied at the museum. His interpretations can even be viewed simply as part of the normal progression of science, in which one paradigm replaces another. As Michael Hammond points out, "the Neanderthal expulsion was but one step in the dismantling of the most influential evolutionary research program in late nineteenth century France."[4] The author of that program, which embodied the ladder form of evolution, had been Gabriel de Mortillet, of whom Boule had long been critical. To be instrumental in overturning what has come to be regarded as an outdated paradigm is every scientist's aim. Boule was no exception.

The reasons Boule took the position he did are thus perhaps understandable. But the manner in which he did it is puzzling. He described the Chapelle-aux-Saints individual as brutish in facial appearance, as having a short, forward-slung neck and a slouching, stooped, knees-bent posture. Moreover, although the individual's brain was at least as large as modern human brains, as is the case with all Neanderthals, Boule concluded that it was poorly developed in those areas which give *Homo sapiens* its great intellectual preeminence. It was indeed the classic caricature of the slow-witted caveman.

All this forms a very consistent picture, and was roundly endorsed by the British establishment. For instance, Elliot Smith, the renowned neurologist, said, "However large the brain may be in *Homo neanderthalensis,* his small prefrontal region is sufficient evidence of his lowly state of intelligence and reason for his failure in the competition with the rest of mankind."[5] Sir Arthur Smith Woodward agreed: "The brain, though great in quantity, may have been low in quality."[6]

What is puzzling about all this is that the anatomy simply does not support the strong conclusions made initially by Boule and secondarily by Elliot Smith, Arthur Keith, Smith Woodward and others. And yet so authoritative was Boule, and so congenial were the conclusions that flowed from his work, that the Chapelle-aux-Saints skeleton in particular and Neanderthal anatomy in general were not seriously reconsidered until the 1950s and '60s. This new research revealed a very different picture.

For instance, Neanderthal brains "do no show 'primitive' features, if size, convolutional patterns, and asymmetries are considered together,"[7] concludes Ralph Holloway, an anthropologist at Columbia University, New York, who is probably the world's expert on modern and fossil hominid brains. Equally, Loring Brace responds point by point to the "specializations" Boule said he recognized in La Chapelle-aux-Saints: "There is no trace of evidence that Neanderthalers had exceptionally divergent great toes or that they were forced to walk orang-like on the outer edge of their feet; there is no evidence that they were unable fully to extend their knee joints; there is no evidence that their spinal columns lacked the convexities necessary for fully erect posture; there is no evidence that the head was slung forward on a peculiarly short and thick neck."[8]

It is true that the Chapelle-aux-Saints individual probably walked with a stooped posture in his later years, but this was a result of extensive arthritis in the spine, a fully evident pathology that Boule mentioned in passing but apparently discounted. "He did not see that this provided the basis for a possible alternative explanation for the remarkable posture of the skeleton,"[9] says Hammond. In other words, Boule was not blind to the evidence, but apparently he was blind to what it implied. Why?

"I have been able to find absolutely no indication, or even a hint, that Boule fudged or fraudulently mis-represented his research," concludes Hammond. "Boule sincerely looked upon science as 'one of the principal sources of happiness' when 'guided by that interior flame,' the 'love of Truth.' His La Chapelle-aux-Saints reconstruction was his most important contribution to this search; and he believed that the best scientific techniques available had been used, and that his conclusions were dictated by the data themselves." It is, in fact, a common fantasy, promulgated mostly by the scientific profession itself, that in the search for objective truth, data dictate conclusions. If this were the case, then each scientist faced with the same data would necessarily reach the same conclusion. But as we've seen earlier and will see again and again, frequently this does not happen. Data are just as often molded to fit preferred conclusions. And the interesting question then becomes "What shapes the preference of an individual or group of researchers?" not "What is the truth?"

It is clear that Boule went beyond the evidence of his eyes—perhaps to press more persuasively his version of the Truth. Michael Hammond suspects that, given the evolutionary model that was prevailing at the turn of the century, a simple, objective description of the robusticity of Neanderthal anatomy might have been inadequate to persuade many anthropologists that the species should be excluded from human ancestry altogether. "Without the stooping carriage, the morphological differences between the Neanderthals and modern man would not have been sufficient to so definitively expel the Neanderthals from a place in the evolutionary origin of man," guesses Hammond. To ensure expulsion, Boule required Neanderthals to display a distinctly apelike, stooping gait and many other "primitive" characteristics; he would exaggerate the differences from modern humans and minimize the similarities. Boule may have realized—consciously or subconsciously—that he had to overstate his case in order to ensure that it would be noticed. Such a tactic is common in the intellectual development and promulgation of new ideas, and for Boule it worked handsomely. His commitment to viewing human history as a bush, not a ladder, with Neanderthal as one of its side branches, had been fulfilled.

The scientific literature of Boule's time is replete with expression of Edwardian revulsion and even moral indignation at the supposed brutishness of Neanderthals. It would, however, be a mistake in Boule's case to conclude that his technical assessments were based on this perceived brutishness. In fact, it was rather the other way around. His preconceptions—primarily that human history was like a bush, not a ladder—demanded that Neanderthals be as different as possible from modern humans, and so he needed to exaggerate those differences which did exist and even invent some which didn't. The result was that Neanderthal looked more brutish than he really was.

For half a century Boule's legacy from La Chapelle-aux-Saints was the keystone to most, but not all, anthropological views about Neanderthal Man: he was a cousin, not a brother, *Homo neanderthalensis* to our *Homo sapiens*. Then, from the mid 1950s through the 1970s, Loring Brace and others led a revival of the notion that Neanderthal was indeed our direct ancestor, and not a side branch that became extinct. As a result, most anthropologists agree that Neanderthal should be called *Homo sapiens neanderthalensis*—in other words a subspecies and a very close relative to our own. (But, as so often happens in the science of paleoanthropology, the intellectual cycle is due to turn once again, and currently there is a gathering sentiment in favor of pushing La Chapelle-aux-Saints and his fellows onto a side branch once more. The reasons for doing so now are different from Boule's seventy years ago, so it seems he may have been right after all, but for the wrong reasons.)

When Boule's work on La Chapelle-aux-Saints was approaching its grand culmination in 1912, with major publications in the *Annales de Paléontologie*, he had been faced with something of a problem, as Michael Hammond explains. "The definitiveness of Boule's conclusion on the Neanderthals suffered from the fact that a great gap remained in the paleontological record. At the end of the first decade of this century, there existed no clear evidence for a pre-Neanderthal population with specializations significantly different in their pattern of development than those of the Neanderthals, and which could be taken as an indication of a more sapiens-like line of evolutionary change. Obviously, if the Neanderthals were not ancestral to man, there must have existed other populations undergoing other evolutionary developments."[10] In other words, Boule's conclusions provided a clear prediction, which needed to be confirmed by the discovery of the right kind of fossils if his line of argument was to carry weight. "It was precisely at this time that the Piltdown Man emerged with its saintly human forehead lacking the great [browridge] of the Neanderthals."[11] At one stroke, the gap was filled.

So it was that Neanderthal Man and Piltdown Man became paleontological partners, each one requiring the other for its existence. "A common theme among anthropologists involved in the reconstruction and defence of Piltdown was the recent dismissal of the European Neanderthals from human ancestry," says Hammond. "Indeed, the most famous anthropological forgery [Piltdown] of this century was firmly rooted in one of the most influential anthropological mischaracterizations [La Chapelle] of the century." Sir Arthur Smith Woodward brought the two together nicely in his announcement of Piltdown in December 1912. The discovery, he said, "tends to support the theory that [Neanderthal Man] was a degenerate offshoot of early man, and probably became extinct; while surviving man may have arisen directly from the primitive sources of which the Piltdown skull provides the first discovered evidence."[12]

Given all the many anatomical incongruities in the Piltdown remains, which of course are glaringly obvious from the vantage of the present, it is truly astonishing that the forgery was so eagerly embraced, by much of the British establishment at least, and by some notable North American anthropologists, including Henry Fairfield Osborn. The forgery was perfectly tailored, not technically but theoretically, and in the timing of the series of discoveries too. For instance, the first discoveries announced included parts of the obviously humanlike cranium and the equally obviously apelike jaw. But there was no canine tooth, which was a subject of some considerable interest because of the unusual wear pattern it might bear. Sir Arthur Smith Woodward predicted publicly what he thought such a tooth would look like, and within a few months one was found. His prediction was vindicated to the finest detail.

Later, in 1917, when some opinions were still wavering, another discovery was made, this time 3 kilometers from the original site. The arrival of Piltdown 2, as it was called, served to still the doubts of many observers, because it seemed to show that the original had been no freak. "There is no longer any reason for attaching any significance to criticisms that have been made by anthropologists not acquainted at first hand with all the evidence now available,"[13] said Elliot Smith of the import of this second Piltdown Man.

"If there is a Providence hanging over the affairs of prehistoric men, it certainly manifested itself in this case," joined in Henry Fairfield Osborn. "The three minute fragments from this second Piltdown man . . . are exactly those which we should have selected to confirm the comparison with the original type."[14] The notion that the skull and jaw had belonged separately to a human and an ape had indeed been widespread. "It would have been very difficult to dislodge this opinion, so widely entertained in Europe and America, but for the overwhelming confirmation afforded to Smith Woodward by the discovery . . . of a second Piltdown man."

Indeed, this second discovery almost broke down the skepticism of Marcellin Boule, who believed that although the skull represented a type of Dawn Man of the sort he had predicted must exist, the jaw must have belonged to an ape. "In the face of these new facts, I am not inclined to be as positive as formerly," admitted Boule. "But I must add that my doubts have not been completely laid to rest."[15] The Germans, however, for the most part remained unmoved in their disbelief. Nationality was indeed a strong predictor of an individual anthropologist's position on Piltdown, as indeed it had been with Neanderthal.

One reason that Britain offered such fertile soil for planting the forgery was that most of the theoretical developments that shaped it had originated there. As mentioned earlier, in addition to the expulsion of the Neanderthals, there were three major issues: first, Arthur Keith's conviction about the great antiquity of modern forms of man; second, William Sollas' work on mosaic evolution; and third, Elliot Smith's theories on brain expansion's leading the way in human evolution.

It is something of a British anthropological tradition that modern forms of man originated deep in geological history. Arthur Keith was the principal proponent in his day, and it is more than a little interesting that Louis Leakey, who for a time was a close associate of Keith's, carried on the tradition. Like Osborn, Keith considered the human brain to be so special that only a very long period of slow evolution could have fashioned it from a more primitive state. As mentioned earlier, his obsession with the idea had led him erroneously to accept two modern skeletons, Galley Hill Man and Ipswich Man, as being of ancient origin. When Piltdown Man came along, once more it seemed to offer evidence in support of his cherished theory. "By 1912, Keith was definitely looking for evidence in this regard, and was obviously ready to suspend much critical judgement on almost any fossil which gave more weight to his idea,"[16] says Michael Hammond.

William Sollas, an anthropologist from Oxford University, actually came closer than anyone else to predicting the form of Piltdown before it turned up. Indeed, in 1910 he had said a human ancestor with a large brain and an apelike jaw was "an almost necessary stage in the course of human development." Until Sollas began working on the idea of mosaic evolution, in which different parts of an organism might evolve at different rates, the whole process was seen as rather regular. At the turn of the century the idea was not only that human evolution followed a straight and regular progression, but also that the body "would at each stage become a little less ape-like, a little more human-like," as Keith put it. If this notion had still prevailed in 1912, the Piltdown forgery would never have been accepted as genuine. More to the point, it almost certainly would not have been perpetrated in the form in which it was. Sollas' influence in arguing that evolution can work at different rates in different parts of the body made plausible the combination of a human cranium with an apelike jaw.

Keith eagerly embraced Sollas' arguments, and saw them expressed in Dubois's find from Java as well as in Piltdown. "The same irregularity of expression of parts is evident in the anatomy of *Pithecanthropus*, the oldest and most primitive form of humanity so far discovered. The thigh bone might easily be that of a modern man, the skull cap that of an ape, but the brain within that cap, as we now know, had passed well beyond an anthropoid status," observed Keith. Mosaic evolution was seen in Java Man and in Piltdown Man, just as it should be—or so it seemed.

Perhaps the most enthusiastic promoter of Piltdown Man was Elliot Smith, for whom it represented something of an intellectual triumph. "In the two years before Piltdown's emergence, Elliot Smith was bringing to conclusion a decade of study on the evolution of the brain in man and other primates," explains Michael Hammond. "He decided that the growth of the brain was the fundamental factor in man's evolution. . . . By 1912 he would be searching for fossil evidence to support his theory on the precocious modernization of the brain and would be tempted to accept any fossil which indicated the leading role of the brain in human evolution."[17] Indeed, when the Piltdown "fossil" eventually came along, Elliot Smith said that "the outstanding interest of the Piltdown skull is the confirmation it affords of the view that in the evolution of man the brain led the way."[18]

So it was that with its human cranium, ape's jaw, and supposed early Pleistocene origin, "No other morphological combination could have fulfilled the theoretical ideas of scientists such as

Boule, Keith, Elliot Smith and Sollas," says Hammond.[19] It was a jewel of a forgery. Robert Broom accepted it. So, briefly, did Louis Leakey. And Henry Fairfield Osborn considered it to be the embodiment of his Dawn Man theory. But even those who did accept Piltdown as a genuine fossil typically pushed it out on a side branch, extinct without issue. Nevertheless, its existence "proved" that there were forms of early, primitive men evolving in this most precocious manner.

The real interest of Piltdown, however, is not so much where on the family tree—or bush—it was hung, but how those who believed in the fossil saw in it what they wanted to see. Remember, the cranial parts were those of modern man, *Homo sapiens,* an individual who died perhaps 2,000 years ago at most. And the jaw is that of a modern orangutan, chemically treated to make it look like a fossil and with cheek teeth filed down to make them look humanlike. Given this mischievous chimera, this is what was said of them.

"The Piltdown skull, when properly reconstructed, is found to possess strongly simian peculiarities," noted Elliot Smith. "In respect of these features it harmonizes completely with the jaw, the simian form of which has not only been admitted, but also exaggerated by most writers."[20] In other words, Elliot Smith was able to see signs of humanity in the orang jaw and features of an ape in the human cranium. "That the jaw and cranial fragments . . . belonged to the same creature there has never been any doubt on the part of those who have seriously studied the matter,"[21] he opined somewhat peremptorily in 1914. It wasn't that the chances were minuscule that a human and an ape might lie down dead cheek to cheek in England's distant Pleistocene. For Elliot Smith, the anatomy unequivocally showed that Dawn Man sported an apelike jaw, just as you would expect him to.

Being a neurologist, Elliot Smith also examined the form of the brain impressed on the inner surface of the cranium. "There are clear indications," he said, "that mere volume is not the only criterion of mental superiority. Those parts of the organ which develop latest in ourselves were singularly defective in [Piltdown]."[22] There are clear echoes here of Boule's assessment of the Neanderthal's supposed inferior mental capabilities, simply because of an assumed primitiveness. Remember, Elliot Smith was in fact describing a fully modern human brain.

Soon after the first Piltdown material was recovered, Sir Arthur

Smith Woodward, who named the fossil *Eoanthropus dawsonii*, reconstructed the cranium. Lacking large segments of the anatomical jigsaw puzzle, Smith Woodward had to make some guesses as to how the pieces he had might relate to each other. Apparently misidentifying some minor anatomical landmarks on the interior of the cranium, he assembled a skull that not only was erroneously small (just over 1,000 cubic centimeters) but also appeared to have certain primitive anatomical features. This reconstruction deeply impressed Elliot Smith. Sir Arthur Keith, however, challenged the accuracy of the reconstruction and did one of his own, eschewing the errors Smith Woodward had committed. Keith's version not only was much bigger (about 1,500 cubic centimeters), but also lacked the primitive features erroneously present in Smith Woodward's. Quite an intellectual battle ensued over who was correct, and at one point Keith offered to demonstrate his skills in anatomical reconstruction. He had to attempt a cranial reconstruction using just a few fragments of a modern skull whose shape and size were known and which had been broken for the purpose. Keith showed himself to be a match for the task, but that still did not settle things.

"I regret to record that it most unfortunately gave rise to a rather acrimonious, and somewhat painful, controversy between Keith and Elliot Smith," commented Sir Wilfred Le Gros Clark, who was instrumental in exposing Piltdown as a forgery in the fall of 1953. "Why . . . did not [Keith's] correction immediately raise suspicions of the authenticity of the Piltdown fossils?" asked Le Gros Clark. "Because of its personal nature the controversy [between Keith and Elliot Smith] certainly clouded the issues and befogged the atmosphere of scientific discussion. Elliot Smith had failed to recognize the true significance of Keith's correction, and, in spite of it, still maintained that the skull and brain did show markedly primitive and simian characters, while the ape-like character of the jaw, in his view, had been exaggerated. In his day Elliot Smith's authority carried greater weight (and rightly so, for he was a very eminent anatomist), so that not only did he persuade himself that his original interpretation of the skull and endocranial cast had been fundamentally right, he also seems to have persuaded biologists in general that this was so."[23]

But in spite of their differences of opinion, both Keith and Elliot Smith continued to accept Piltdown Man as a vindication of their own ideas, each for his own different reasons. Keith, who viewed the skull as essentially modern in form, saw it as a confirmation of the antiquity of modern types of man. At the same time, Elliot Smith claimed the cranium to be distinctly primitive in form and to prove that "in the evolution of man, the brain led the way." Built from a set of plausible theoretical constructs, the "protective screen" surrounding Piltdown proved to be unusually resilient. "All the collateral lines of evidence appeared to be mutually confirmatory and in complete harmony with each other," commented Le Gros Clark when he discussed the forgery at a lecture at Britain's Royal Institution not long after its exposure. "So much so, indeed, that . . . none of the experts concerned were led to examine their own evidence as critically as otherwise they would have done." A clearer message for the process of science can hardly be imagined.

During the four decades following the discovery of Piltdown, the influence of the British school began to decline. Other fossil discoveries were being made in both Asia and Africa, and Piltdown came to look more and more enigmatic. "A puzzle of a most bewildering kind"[24] was how Le Gros Clark described it in 1950. "By this time Piltdown just didn't make any sense," recalls Sherwood Washburn. "I remember writing a paper on human evolution in 1944, and I simply left Piltdown out. You could make sense of human evolution if you didn't try to put Piltdown into it."[25] Earnest Hooton, Washburn's mentor and one of the most ardent Piltdown supporters in the United States, was outraged by Washburn's paper. "You can't leave out the evidence," Hooton stormed at his former pupil.

But for most people, Piltdown eventually ceased to be evidence at all. It was gradually pushed to one side, awaiting some kind of resolution, though no one ever guessed that forgery would be the answer. Gerrit Miller, of the Smithsonian Museum of Natural History in Washington D.C., came closest when, in 1915, he made the following comment: "deliberate malice could hardly have been more successful than the hazards of deposition in so breaking the fossils as to give free scope to individual judgment in fitting the parts together."[26] For Miller, this was merely a rhetorical observation, not a serious conjecture. But forty years later, it proved to be uncannily accurate.

Appropriately enough, in addition to being involved in the exposure of the Piltdown fraud, Le Gros Clark was also instrumental in rescuing the Taung skull from anthropological oblivion, an event that occurred in 1947. At the end of the previous year Le Gros Clark had left his Oxford laboratory for an extensive African visit, beginning in Johannesburg. In addition to the Taung skull, he was able to examine at first hand the many similar specimens that had been recovered since 1936, which is when Dart's friend and associate Robert Broom had found the first australopithecine fossil since Dart's original discovery.

"I had first made careful examination and taken copious notes of the skulls and teeth of over a hundred anthropoid apes in various museum collections in order to reassure myself of the normal variation that they show in their anatomical features and to compare them with the descriptions of the australopithecines already published by Dart and Broom," Le Gros Clark later recalled. "I still remained doubtful of the suggestion that the australopithecines were hominids rather than pongids [that is, apes], and I still inclined to take the latter view. So far as the contentions of Dart and Broom were concerned, therefore, I felt I was going to South Africa as the 'devil's advocate' in opposition to their claims."[27] After two intensive weeks of poring over the fossils and visiting the cave sites near Johannesburg, Le Gros Clark's reservations dissolved. "The results of my studies were very illuminating," he noted. "They at last convinced me that Dart and Broom were essentially right in their assessment of the significance of the australopithecines as the probable precursors of more advanced types of [humans]."

The overall form of cranium, the shape of the face, the detailed anatomy of the teeth and very humanlike architecture of the limb fragments found in the Transvaal caves were sufficient to convince the devil's advocate that here was a creature which, though primitive, was distinctly allied to the human family. It was indeed a hominid, not a pongid, Le Gros Clark concluded.

The anthropological community soon heard of Le Gros Clark's assessment, because early in January 1947 he traveled from Johannesburg north to Nairobi, Kenya, where he presented his findings at the First Pan-African Congress on Prehistory, which was organized by Louis Leakey. In his paper Le Gros Clark formally referred to the australopithecines as hominids, rather than by some vague term associating them with apes, as had become common practice. This wasn't the first time the term "hominid" had been applied to Taung and its fellows, but it was the first time a practitioner of the authority of Le Gros Clark had done so, and he did it in such a manner that it could no longer be ignored.

A report by an anonymous correspondent in the editorial pages of the 15 February 1947 issue of *Nature* soon carried Le Gros Clark's message worldwide. "The suggestion that the Australopithecinae are to be regarded as anthropoid apes . . . must almost certainly be ruled out," reported the correspondent. "There appeared no room for doubt that Dart and Broom had certainly not over-estimated the significance of the Australopithecinae, and their interpretations of these fossil remains were entirely correct in all essential details."

The very same day that this issue of *Nature* was published, Sir Arthur Keith went to his study, at his house in the village of Downe in Kent, and wrote a short letter to the journal's editor. "I was one of those who took the point of view that when the adult form [of *Australopithecus*] was discovered it would prove to be near akin to the living African anthropoids—the gorilla and chimpanzee," he wrote. "I am now convinced . . . that Prof. Dart was right and that I was wrong." A prompter and more thorough capitulation could hardly be imagined. This acknowledgment came exactly twenty-two years after Keith had publicly declared that the Taung baby was probably an ape, in the 14 February 1925 issue of *Nature*. Keith did, however, complain that the name *Australopithecus* was long and unwieldly, and suggested that the australopithecines should be called "Dartians" instead. The idea never caught on.

In the twenty-two years between the announcement of the Taung fossil and the beginning of public recognition following Le Gros Clark's pronouncement, Dart had more or less turned his back on anthropology. He had done so partly out of dismay at the trenchancy of the opposition to his ideas, but also because he was plentifully occupied with other tasks in his department and in the university in general. It wasn't until 1945 that Dart once again took an active interest in anthropological fieldwork. Had it not been for the energy and enthusiasm of Robert Broom, therefore, Le Gros Clark might not have had any reason to travel to South Africa at all, for it was Broom who in 1933 instigated searches in new limestone caves in the Transvaal, the Taung cave having been quickly obliterated by mining activities soon after the child's skull was recovered.

Virtually alone, Broom had been Dart's supporter from the start. Two weeks after Dart's original *Nature* paper was published in February 1925, Broom had gone to see Dart at the medical school. "When Broom came into the room he walked straight past me, straight past some of my staff who were with me and immediately dropped to his knees in front of the Taung fossil,"[28] recalls Dart. "Broom said it was 'in adoration of our ancestors.' It was a remarkable occasion. I was very surprised."

Broom, a medical man, was an extraordinary individual, who in 1933, at the age of sixty-seven, launched himself into a new career, that of paleoanthropologist. Three years later he had discovered the first australopithecine since the announcement of Taung. More important, this one was adult.

Quite legitimately, some of the reservations expressed about the interpretation of the Taung skull had been that it was a juvenile. Not only does the anatomy of juveniles alter as they become more mature, thus making interpretations somewhat tricky, but the juvenile apes have a very humanlike form. Dart believed he could detect incipient humanity in the Taung child, but others considered that he was simply being misled by the transient humanlike features of a young ape. So when Broom found part of a cranium and associated brain cast of a mature adult, there should have been no more reason to balk at Dart's original proposals. Except that the anthropological community was mesmerized by the series of remarkable finds coming out of the Chou Kou Tien cave, near Peking, the famous Peking Man fossils which tragically were lost near the beginning of the Second World War.

Even the intellectual weight of William King Gregory, who had by this time thrown his support behind Dart, proved to be ineffective against prevailing professional sentiment. In 1930 he had written in *Science*, "If *Australopithecus* is not literally a missing link between an older dryopithecoid group and primitive man, what conceivable combination of ape and human characters would ever be admitted as such? . . . *Australopithecus*, to judge from its skull and dental characters, was a pioneer in a new line, as held from the first by Dart."[29] This, remember, was at a time when Henry Fairfield Osborn, Gregory's boss, was extolling the nobility and antiquity of Dawn Man, who most definitely was no relation to an ape.

A few years later, in 1938, Gregory had an opportunity to visit Dart, Broom, and their fossils in South Africa, and became even more convinced of the validity of Dart's views. Taung was "the missing link no longer missing," he told an audience of the Associated Scientific and Technical Societies of South Africa. "Dr. Dart concluded at that time that his form represented a long step in the direction of the human race; and I do not believe," concluded Gregory, "after the most critical studies my colleagues and I have been able to make, that any reasonable exception can be taken to that conclusion." This point "marked the beginning (but only the beginning) of the scientific reassessment of the place of *Australopithecus* in human evolution,"[30] notes Charles Reed, who has made a study of the history of this period.

Broom continued to amass more adult Taung-like fossils, and just two years after his first find he came across what was obviously a second type of australopithecine. This one was a much more robust version of the Taung type, and, appropriately enough, has been called *Australopithecus robustus*. In fact, Broom was one of the most egregious of "splitters," christening virtually each new fossil with a new scientific name, so that for a time the hominids from the South African caves were a veritable zoo of names. They have since been rationalized to the two forms of *Australopithecus*: *africanus* and *robustus*. The discovery of a second species of *Australopithecus* was totally unexpected, however, and it began what eventually was to become a long debate about just how "bushy" the human family tree really is.

Although Broom was enthusiastically searching for australopithecines, he was also a believer in the Piltdown Man, which caused him some problems. At one point it looked as if Piltdown were older than the australopithecines, which implied to Broom that, according to Reed, the "australopithecines could not have been ancestral to 'men' but were instead relatively unchanged descendants of those ancestors." Nevertheless, Broom argued that the stock from which the australopithecines of the Transvaal derived was older than Piltdown and therefore could in theory be ancestral to it. But it was only a guess, and his thinking on the matter was always somewhat confusing.

"Broom continued to use the words 'ape' or 'ape-men' in referring to the australopithecines," says Reed. And so did Dart. In his 1925 *Nature* paper Dart had plainly referred to Taung as an ape: "an extinct race of apes intermediate between living anthropoids and man." Even in 1967 "Dart was still calling australopithecines 'man apes' and pithecanthropines [that is, Peking Man, Java Man, etc.] 'ape-men,' " Reed observes. In fact, Dart's equivocation over what the Taung child should be called—hominid or ape—undoubtedly contributed to the anthropological community's tardiness in welcoming *Australopithecus* into the human family.

Given the rampant pithecophobia of the time, it was perhaps unwise of Dart and Broom to use the word "ape" at all in connection with creatures they clearly saw as putative members of the human family. "Even the scientific community was not then ready for early Pleistocene or late Pliocene [about 2 million years ago] hominids that didn't yet look like they thought hominids should," observes Reed. "Not only was the large-brained, ape-jawed Piltdown specimen a model of what 'early man' should be, but many physical anthropologists of the time expected hominid ancestors as primitive as the australopithecines were to have lived much earlier, since they considered the division between hominids and other primates to have occurred in the Oligocene [at least 25 million years ago] or earlier (thus avoiding relationship with the apes), a view which has lingered on."

Le Gros Clark makes a similar point, ironically in a book titled *Man-Apes or Ape-Men?* "In all these early discussions we made the mistake of using the colloquial term 'ape' without defining exactly what was meant by it," he said. "In some later controversies, also, the same loose usage of the term led to a good deal of confusion. Such confusion could have been avoided if we had used instead the scientific terms of zoological classification, Pongidae (instead of 'apes') and Hominidae (instead of 'man')."[31] Readers may judge for themselves whether semantic massaging of this sort would have eased the very obvious pain of embracing within the human family something that looked so very much like an ape.

Dart's differences with Keith over the implications of Taung were superficially matters of emphasis, but they proved to be crucial. "Keith considered *Australopithecus* to have been an extinct side-line of apes, showing some few human-like evolutionary tendancies but still within the general group of chimpanzees and gorillas," observes Reed. "From Dart's point of view, Keith interpreted the evidence negatively, whereas from Keith's point of view Dart had over-emphasized what seemed to be a relatively few hominid-like characters and had tended to weight too lightly all the pongid characters." Reed suggests that Keith's major error was in comparing the anatomy of the Taung child with that of modern children. "Naturally, he found major differences, which he used to discount the possibility of *Australopithecus* being an early hominid ancestor." This type of error has been repeated—consciously or unconsciously—many times in the history of paleoanthropology, and many times in particular in relation to the status of the Taung child and his fellows.

For instance, Solly Zuckerman appears to have fallen into this trap while championing a last-ditch stand against the acceptance of *Australopithecus* as hominid during the 1950s and '60s. Zuckerman, ironically enough a South African, emigrated to England in 1926 and by turns was knighted and then further ennobled, to become Lord Zuckerman. His Lordship eventually became scientific adviser to the highest levels of government, and perhaps the most powerful voice in British science. He was also, apparently, in William King Gregory's terms, a pithecophobe.

Zuckerman was a medical student in Cape Town in the early 1920s, and, by chance, had as one of his final examiners in anatomy Professor Raymond Dart. Dart was impressed by Zuckerman, and gave him a letter of introduction to Grafton Elliot Smith. "It was probably the worst thing I could have done for my own future,"[32] Dart now laments. So it was that Dart, the discoverer and promoter of *Australopithecus*, who had felt expelled from the London scene of Elliot Smith, Keith and others, helped Zuckerman make the reverse journey, where he would become a strident critic of *Australopithecus*.

Zuckerman insisted that using conventional descriptive methods of anatomy, anthropologists simply had not made the case that *Australopithecus* was a hominid. He argued that the only true way to do so was in careful measurement and statistical analysis, by which method he concluded again and again that the australopithecines were more ape than human. Therefore they were not hominids. Zuckerman was committing Keith's error. Zuckerman was demanding that in order to be accepted as members of the human family, the australopithecines must cross the dividing line that separates ape from human. Apparently, anything less than 50 percent human would not do. "The most sophisticated techniques will only give dusty answers to badly designed questions" was how one authority characterized Zuckerman's efforts.[33]

Being in South Africa as a young man, Zuckerman had an opportunity to examine the Taung skull for himself. "He determined to his own satisfaction that *Australopithecus africanus* was an ape," notes Reed, which conclusion Zuckerman published in 1928. "He has not yet seen sufficient evidence to change his mind."[34] Zuckerman, like Sir Arthur Keith before him, believes that apes and humans diverged way back in the Oligocene, some 25 million years and more ago, a view he developed early in his career and clings to still. It is therefore difficult to see what could persuade him to accept as hominid anything that was anatomically primitive and yet lived only a couple of million years ago. To be admitted into the human family, a creature as recent in time as 2 million years must surely be much more humanlike and much less apelike than *Australopithecus* obviously was, for in Zuckerman's estimate it would have been separated from the apes for at least 20 million years.

Zuckerman worked for a time in Le Gros Clark's department at Oxford University—a professional relationship that ruptured somewhat violently following Clark's 1947 pronouncements on *Australopithecus.* Following his departure from Oxford for Birmingham University, Zuckerman promoted an active research program in the application of metrical methods in paleoanthropology. The results obtained by Zuckerman and his students were not received with enthusiasm. He commented wryly on this in 1966, saying, "It is something of a record for an active team of research workers whose strength has seldom been below four, never to have produced an acceptable finding in 15 years of assiduous study."[35]

In spite of Zuckerman's lack of success in persuading the paleoanthropological community of its errors, Dart clearly remains bitter about the man whom he introduced to Elliot Smith. The reason is that Dart's most important scientific document about the Taung child—a major monograph he completed in 1930—has never been published, a state of affairs he attributes to Zuckerman's influence. It is true that Zuckerman very quickly became a respected and prominent member of Elliot Smith's circle of colleagues. And it is true that when Dart was in London in early February 1931 to discuss publication of the monograph by the Royal Society he dined at Elliot Smith's house, with Zuckerman present. But, as is the nature of these things, there is no clear evidence one way or the other why the monograph was rejected by the Royal Society, and Zuckerman dismisses Dart's suggestion as nonsense. Dart remains convinced, however, and said recently, "When they refused to publish my monograph, I began to be suspicious of Zuckerman's influence." Hence his acerbic comment at the Taung diamond-jubilee celebrations in February 1985: "I just wish that man Zuckerman could be here to see all this." Zuckerman, now retired, was not invited to the Johannesburg meeting, nor, in all probability, would he have attended had he been.

The Johannesburg gathering was indeed an impressive demonstration of the revolution "wrought by Dart's brilliant insight," as Phillip Tobias put it. Long gone was the specter of Piltdown. Long gone was the notion—in reality, a cherished hope—of a large-brained ancient ancestor. Long gone was the full-blown pithecophobia of Osborn and his ilk. Taung and his fellows were small-brained and distinctly apelike. They were, in geological terms, recent creatures of the African plains. And yet—Zuckerman and his colleagues aside—they were universally accepted as members of the human family. They were hominids. Dart had been right about the Taung child.

The paleoanthropological community's acceptance of Le Gros Clark's conclusions in 1947 had been rapid and complete, as most scientific revolutions tend to be. It's true that residues of pithecophobia persisted, so that australopithecines have sometimes been depicted as being more apelike in their style of locomotion than the early members of the genus *Homo*. Even as recently as the 1970s, professional anthropologists would assume, on the flimsiest of evidence or more usually on no evidence at all, that the australopithecines' bipedalism was an ungainly, energetically inefficient "stagger" not unlike that of a gibbon as it moves over open ground. Some even assumed they used their knuckles for support, as modern chimpanzees and gorillas do. In fact, proper anatomical analysis has shown australopithecines' locomotion to have been perhaps even more efficient than that of *Homo sapiens.* Here again was an example of that old problem of differences being exaggerated while similarities were minimized. From the neck up, australopithecines certainly looked a lot like apes, and so people were too easily led to accept that they were apelike in other ways too. But no mainstream paleoanthropologist wanted to go so far as to kick Taung and his fellows out of the human family.

What researchers have principally been pondering in these past four decades is precisely where within the human family *Australopithecus* fits. Was it a direct ancestor of the *Homo* lineage, which eventually finished up as modern man? And if so, which species of *Australopithecus* gave rise to *Homo*? Or were the australopithecines merely evolutionary cousins, having derived in parallel with the *Homo* line from some as-yet-undiscovered common ancestor? Louis Leakey, for instance, never accepted the idea that *Homo* had evolved from an australopithecine. And the human family trees that Richard Leakey draws sometimes appear to relegate Taung and his fellows to the status of cousins, not brothers. But most anthropologists today believe that a species of *Australopithecus* once occupied a place in the direct line to modern man.

Paleoanthropological passions had been deep and strongly expressed over Neanderthal and Piltdown, which for so long were obstacles to the acceptance into the human family of Taung and his fellows. Those obstacles are now gone, but the passions remain, this time expressed in the modern debates over the precise status of *Australopithecus* within the human family.

CHAPTER 5

RAMA'S APE: RESURRECTED

"I will never again cling so firmly to one particular evolutionary scheme," announced David Pilbeam at the beginning of 1978. "I have come to believe that many statements we make about the hows and whys of human evolution say as much about us, the paleoanthropologists and the larger society in which we live, as about anything that 'really' happened."[1]

This dramatic public recantation shocked the paleoanthropological profession, because it represented more than a shift in just one person's philosophy of science. For a decade and a half Pilbeam, with his Yale colleague Elwyn Simons, had embodied the science's virtually unanimous commitment to one particular view of human origins. Namely, that humans split away from ape ancestors at least 15 million years ago; and that the first member of the line leading to us was a baboon-sized creature known as *Ramapithecus*.

With Pilbeam's defection, the way was opened for the ascendancy of a rival hypothesis. Specifically, that *Ramapithecus* was not on the human line at all, and that we shared a common ancestor with the apes until as recently as 5 million years ago. This latter idea was anathema to most paleoanthropologists, not least because its proponents were largely outside the profession: they were biochemists, of all things.

Four years after his bold public confession, Pilbeam wrote the following: "I have had doubts about the hominid status of *Ramapithecus* since at least the mid 1970s, although I have avoided stating explicitly that it was *not* a hominid until I had a clearer idea of what it was! I now have a clearer view."[2] That clearer view stemmed from the analysis of a remarkable fossil face of an extinct ape found in Pakistan in 1980. The results of the analysis, says Pilbeam, "triggered a series of insights that have crystallised new ideas for me." *Ramapithecus*, Pilbeam concluded, is not a proto-human at all, but is somehow related to the ancestors of the modern Asian great ape, the orangutan. These developments encouraged British paleoanthropologist Bernard Campbell to write the following in the 1985 edition of his textbook *Human Evolution*: "Today even the most prejudiced paleontologist, who had little time for a consideration of the biochemical evidence . . . must now agree that the common ancestor of gorilla, chimpanzee and human may not be older than 6 million years." The previous —1974—edition of the book had, like all such texts, proclaimed the split to be three times as old, with little mention of the bio-chemically deduced alternative.

The dethroning of *Ramapithecus*—from putative first human in 1961 to extinct relative of the orangutan in 1982—is one of the most fascinating, and bitter, sagas in the search for human origins. Some practitioners describe it as an exemplary illustration of how the science should proceed: that is by changing its hypotheses on the basis of each new item of evidence. Others, by contrast, charge that there are echoes of the Piltdown affair, in the sense of experts seeing in the fossils precisely what they want to see. There is no doubt, however, that in addition to the confection of egos and reputations that spices any academic fight, the controversy over *Ramapithecus* shows once again how the great difficulty of inferring relationships from fossil shapes can lead to serious intellectual battles. But in this case there was an additional component: the claim by biochemists that the study of molecules from living animals is a superior methodology for understanding relationships between humans and apes. No profession takes kindly to the suggestion that its methods for performing its primary task are inferior to those from another, completely unrelated, profession. Paleoanthropologists are no exception.

The story of the rise and fall of *Ramapithecus* as putative First Hominid, and the concomitant ebb and flow of egos and emotions in the body paleoanthropological, is best told in two parts. In this chapter is recounted the enthusiasm with which meager fossil evidence was interpreted—or more properly, overinterpreted—as a sure sign of incipient humanity, both anatomically and behaviorally. The story of the utter collapse of these ideas, in the face of both molecular and fossil evidence, is told in the following chapter.

The two-decade-long reign of *Ramapithecus* as putative first hominid began in November 1961 when Elwyn Simons published a short paper from the Peabody Museum at Yale. The fossil remains of *Ramapithecus* were—and still are—modest, at the time consisting principally of two parts of a single broken upper jaw. As is usually the case, several pieces of the jaw were missing, including most of the palate itself and part of the front, which turned out to be rather crucial. In spite of its fragmentary nature, Simons was able to detect several features that enabled him to infer a relationship with humanlike anatomy. For instance, the canine teeth were small, which contrasts with the long, often daggerlike canines of apes. The individual cheek teeth, the molars, were very like those in modern humans. The face was short and did not protrude as it does in apes. But most important of all, perhaps, the dental arcade as reconstructed by Simons "can be determined as parabolic [like an arch] and not U-shaped."[3] which again is a humanlike as against an apelike characteristic.

Simons said in his 1961 paper that on the basis of these features, *Ramapithecus* "can be defended as being within, or near, the population ancestral to Pleistocene and subsequent hominids." This conclusion was strengthened and extended in a series of a dozen or so papers over the next decade, many of which were coauthored with David Pilbeam, who joined Simons at Yale as a graduate student in the fall of 1963. Simons and Pilbeam suggested that in addition to the hominidlike features in the upper jaw, *Ramapithecus* probably walked about on two legs, not four; used tools to prepare its food; hunted, and had a social life more complex than any ape. It was a powerful package, and most of the paleoanthropological establishment quickly embraced it. Raymond Dart wrote Simons a letter congratulating him on his achievement. Clearly, it was a hypothesis whose time had come.

This ready acceptance of *Ramapithecus* as a hominid was in fact a resurrection of an old idea. The original fossils had been discovered in 1932 by G. Edward Lewis, a doctoral student, during a Yale

University expedition to the Siwalik Hills in India. The name he chose for the fossils, *Ramapithecus*, means "Rama's ape," Rama being a mythical Hindu prince. And the interpretation he offered of them in a 1934 publication was very much along the same line of argument as Simons' was to be, almost three decades later. Lewis' claims, however, were more or less flatly rejected. The reasons for the different response are interesting.

On the face of it, the most obvious reason appears to be a highly critical review by one of the leading paleoanthropologists of the time, Ales Hrdlicka. An immigrant from Czechoslovakia, Hrdlicka had quickly become a leading figure in American science, not least because he founded the American Society of Physical Anthropology in 1930. He was also for many years editor of the society's journal, from which position he wielded substantial power over what was acceptable to the establishment and what was not. However, he chose to damn *Ramapithecus* in the pages of the *American Journal of Science*, which is where Lewis had published his claims for the fossil. In six short pages Hrdlicka tore into Lewis' work, accusing the younger man of committing "a series of errors"[4] and reaching an "utterly unjustifiable" conclusion. *Ramapithecus*, he said, was just an ape.

Naturally, Lewis was extremely upset, not least because Hrdlicka had gained access to the *Ramapithecus* fossils—Lewis' fossils—while Lewis was again in the field. Richard Lull, who was Lewis' research supervisor at Yale, had apparently given Hrdlicka permission to go through Lewis' collections. Lull, it seemed, was in awe of the great man. Lewis, however, was not. Hrdlicka, he says, "thought he was the anointed and elect prophet who had been foreordained and chosen to make such discoveries and demolish the work of anyone else."[5] Hrdlicka's paper was somewhat self-contradictory, and, says Simons, "scattered with blunders and naïvetés that a really good professional simply would not have made."[6] "The man didn't know what he was talking about," recalls Lewis. "So I could not take the paper's content seriously, but did take seriously the possibility of his damaging my reputation."[7]

As an attempt to salvage his reputation, Lewis penned "an unhurried and temperate reply." The rebuttal never found the printed page, however, because the editor of the *American Journal of Science*, Lewis' own supervisor, Lull, declined to accept it. "They refused to publish it," says Lewis, "although they admitted that I had written nothing offensive, because they said Hrdlicka was an important man, and I was a young man whose reputation would be damaged . . . inasmuch as the baldly stated facts and courteous comments would make him look like a fool!" Lewis' thesis—which is described by Pilbeam as "a very good piece of work"[8] and by Simons as "the best opinion people could reach at the time"[9] —was never published. Not long afterward, Lewis left Yale and never really made another important contribution to paleoanthropology.

Hrdlicka had good reason to want to discredit Lewis' work, says Frank Spencer, a scholar of this period of the history of paleoanthropology and of Hrdlicka in particular. "It had nothing to do with the shape of the jaw," he suggests. "It had to do with where the jaw came from—namely, the fringes of Central Asia."[10] In Hrdlicka's view, the western part of the Old World was the wellspring of human origins. Everything in his scheme depended upon this, including his ideas on the eventual peopling of the New World. To have the first hominids appearing in the eastern part of the Old World was therefore simply unacceptable. "So he did a hatchet job on Lewis' work," says Spencer.

Simons considered Hrdlicka's paper to have been of a low standard. "Even a casual examination of this paper is sufficient to show that it bears all the evidence of being a controversial and non-objective contribution," he wrote in his 1961 paper.[11] He even described Hrdlicka's approach as "amateurish."[12] "It looked to me like someone coming into something he didn't know much about, with preconceived ideas,"[13] Simons now recalls.

But as Simons points out, the real key to the cool reception received by *Ramapithecus* lies in a short and, with hindsight, ironic passage in Hrdlicka's diatribe: "[*Ramapithecus*], although in the upper denture, in general, nearer to man than any of the Dryopitheci or the *Australopithecus*, cannot . . . be legitimately established as a hominid, that is a form within the direct human ancestry."[14] In other words, he was arguing that even though the upper jaw of *Ramapithecus* is more humanlike than that of *Australopithecus*, it is still not a human ancestor.

In 1935, when Hrdlicka was tongue-lashing Lewis, most paleoanthropologists were still dismissing *Australopithecus* as some kind of ape. To be palatable as a potential human ancestor to Hrdlicka and his contemporaries, a fossil had to be much more like a human than like an ape, as the reception of Piltdown had so clearly revealed. "Hrdlicka was mainly reacting in line with the then accepted outlook," observes Simons. "Not accepting *Australopithecus* as a hominid, they naturally could not accept *Ramapithecus* either."[15] William King Gregory's support of an ape ancestry for human origins made him sympathetic to Lewis' case for *Ramapithecus*, but he did not choose to champion it strongly. For the most part, then, *Ramapithecus*, like *Australopithecus*, was simply too primitive, too much like an ape for most people's tastes and preconceptions.

So, like Raymond Dart's Taung child, Lewis' little prince had to wait for recognition until there was a change in people's ideas on what an early human ancestor should look like. That change depended on the prior welcoming of *Australopithecus* into the human family, which occurred through the mid 1940s and 1950s; and on the right person's coming along to see in Lewis' fossils what he had seen.

When Simons went to Yale, toward the end of 1960, he did so with excellent credentials: two doctorates, one from Princeton on an obscure group of mammals and a second from Oxford University on the lower primates, or prosimians, that had lived in the Eocene (45 to 25 million years ago). The essence of the Princeton thesis was published in *Transactions of the American Philosophical Society*, which, guesses Simons, must have been read by as many as five people throughout the world. "No one could have cared less about the taxonomy of these mammals," jokes Simons, "so it gave me a chance to be objective."[16]

The move to Oxford allowed him to indulge his boyhood interest in human origins, even though the actual thesis work was on fossil prosimians. His thesis adviser was Sir Wilfred Le Gros Clark, who played such a key role in the belated recognition by the British establishment of *Australopithecus* as a hominid and in the unmasking of the Piltdown fraud. Le Gros Clark had developed anatomical guidelines for distinguishing between apes and humans. One key feature was the shape of the dental arcade: in humans it is like an arch, or parabolic, he noted, while in apes it is U-shaped. But Le Gros Clark tried to avoid focusing on single characters, and instead looked for what he called the "total morphological pattern"—a sense of the larger three-dimensional whole. This was the background that Simons took to Yale, and being an accomplished sculptor as well as a careful scientist, he was good at it.

When he arrived at Yale's Peabody Museum, he found a lot of

Lewis' work instruments and other material still around the laboratory, untouched by his predecessor, J. T. Gregory, who thought Lewis might one day return. He saw Lewis' thesis for the first time, and for the first time handled the original *Ramapithecus* fossils, casts of which he had already seen at Princeton. Simons learned nothing new about the jaw from seeing the original rather than the casts, but the realization of how very similar it was to a human jaw was new, and that came quickly. Within a year he had written and published the now famous 1961 paper, which was called "The phyletic position of *Ramapithecus.*" But the resurrection of Rama's ape to hominid status was taking place against the background of a much more serious task: that of sorting out the horrendous mess of fossil apes and putative human ancestors contemporary with *Ramapithecus*. Collectively, apes and members of the human family are known as hominoids.

Ever since the first Miocene (25 to 5 million years ago) hominoid was discovered in 1856, there had been an orgy in the naming of species, almost to the point of giving each new fossil specimen a new species name. "There were about 25 different genera, and twice as many species, of Miocene hominoids," says Simons. "If these different names had been justified, there would have been some important anatomical differences between the various specimens. But they all looked rather similar to one another." People who work with fossils are sometimes known as "splitters" if they are inclined to give two similar fossils different names and "lumpers" if they generally choose to put the two under the same name. Splitting had been rampant among paleoanthropologists. "Some people even admitted that they were giving their fossil a new genus and species name so as to call attention to how important they thought it was. Everyone who had a fossil come into their hands for description wanted it to be something new—perhaps consciously, perhaps unconsciously—for the purposes of self-aggrandizement."

One reason the plethora of names for the fossils persisted for so long, says Pilbeam, is that no single person had worked on them all. "Louis Leakey and Le Gros Clark had done some work in the 1940s and '50s, but it was quite limited. They indicated that there were too many names. Perhaps Elwyn was influenced by his discussions with Le Gros Clark about this."[17]

Anyway, it was a grand paleontological mess, and it is perhaps little wonder that *Ramapithecus* languished unrecognized for so long in such a welter of names that no one could make any sense of it. One factor that slowed progress here, suggested Simons in 1963, is "the idea, expressed by some vertebrate paleontologists, that the evolution of higher primates, and man in particular, is too controversial and confused a subject to be worth much serious attention."[18] But Simons tackled it with vigor. "By the time I arrived at Yale in the fall of 1963 he was well on his way," recalls Pilbeam.[19] "Everything was falling into place very fast." Within a couple of months of their collaboration Pilbeam had written a paper for *American Scientist*, with Simons as coauthor, that essentially established their position on *Ramapithecus* for several years to come. And within a year the unscrambling of the Miocene hominoid mess was virtually complete. Simons and Pilbeam simplified hominoid taxonomy to just a few genera and a handful of species: they were super-lumpers.

"It was a heady time, very exciting," recalls Pilbeam. "Very quickly we had an extremely tidy story." "David very quickly learned everything I thought about these animals," adds Simons. "It was in his nature to be extraordinarily supportive of the position I had held independently before." The two men worked so well together and their ideas were so compatible that their separate identities virtually merged for a while. One graduate student was convinced that there was just one person, not two, and was acutely embarrassed when one day he ventured to the anthropology office and asked to speak with Dr. Simons Pilbeam.

With the great majority of generic and specific names of the Miocene hominoids cast aside by Simons and Pilbeam in favor of just a handful, the whole picture of higher primates seemed much simpler—too simple, as it turned out. "I can't tell you how many times we were congratulated," says Simons. "People were obviously ready for the classification."[20]

Most people, that is. Louis Leakey wasn't, primarily because in the process two important groups of his fossils had been swallowed up into anonymity. The first was a 15-million-year-old upper jaw, called *Kenyapithecus*, from western Kenya, which he considered to be a very early hominid, like *Ramapithecus*. By coincidence, Leakey had found his putative hominid fossils in the same year, 1961, that Simons resurrected Rama's ape to hominiddom. But *Kenyapithecus* was different from *Ramapithecus*, Leakey proclaimed, and, more important, he thought it was older. He called his fossil by the species name *Kenyapithecus wickeri*.

Later he discovered a second species of this same genus, which he christened *Kenyapithecus africanus*, and which was older still: "the oldest member of the family Hominidae,"[21] he proclaimed. And the second group included a 20-to-25-million-year-old skull he thought might represent an ancestor to modern chimps, which he had named *Proconsul.*

Simons and Pilbeam looked at the *Ramapithecus* jaw, looked at the *Kenyapithecus* jaw, and concluded that there was no biological difference between then, even though they were separated by thousands of kilometers in geography and at least a million years in time. By the rules of nomenclature, because the name *Ramapithecus* had been coined first, Leakey's term *Kenyapithecus* was sunk. And *Proconsul,* by the same type of logic, became a species of *Dryopithecus.* Indignant, Leakey described this as an "extreme example of taxonomic lumping."

Pilbeam was soon to feel Leakey's ire at the super-lumping that was threatening the status of the Kenyan fossils. During a small scientific gathering in Chicago in 1964 organized by the Wenner-Gren Foundation, Pilbeam was presenting some of the thinking that he and Simons had been developing for the revision of the Miocene apes. Twenty participants, one of them Leakey, were sitting around a large table. The format of such meetings, which are acknowledged among the profession to be among the scientifically most worthwhile, includes the prior circulation of papers. There is discussion only, no formal lectures. "Elwyn and I had prepared a paper on our as-yet-unpublished revision of the Dryopithecines, in which we said that *Kenyapithecus wickeri* was *Ramapithecus punjabicus* and that *Proconsul* was a *Dryopithecus,*" remembers Pilbeam. "I was making an oral presentation when Leakey jumped up and started to yell at me, telling me that I was out of order and that 'we don't have to listen to all this again.' I looked around, but no one offered any help. The chairman didn't say anything. So I said to Leakey, 'I'm not going to shut up. I have the floor. You are out of order. *You* sit down and shut up.' Remarkably, he did."[22] Pilbeam finished his presentation and, when the session was over, slunk off to lick his wounds.

"We all thought it was very unfair," says Simons. "David was young in the profession and Louis was a senior member. It is because of things like that that people were prepared to be not very polite to him in print. It wasn't just the blunders he made that encouraged people to criticize him. It was this kind of blustering and arm-waving." In retrospect, Pilbeam reckons that the incident gave him a leg up, because he didn't give in to Leakey. In any case, it set the seal to a more or less continuous animosity between the two men, each taking any opportunity that came along to needle the other in print. There were numerous such opportunities.

Meanwhile, the new career of *Ramapithecus* was well under way. But there was one great curiosity in the fossil collections at Yale. All the fossils that had been allocated to *Ramapithecus* were parts of upper jaws. There were no lower jaws to be found. As lower jaws are by far the most resilient part of the skeleton, being thick, concentrated bone, the discrepancy was difficult to explain. Equally, there were fragments of another creature that Lewis had collected in India, which he named *Bramapithecus* and had considered might be a hominid too. The *Bramapithecus* fossils were all lower jaws. The obvious had happened: an example of super-splitting. The *Ramapithecus* fossils matched the *Bramapithecus* fossils perfectly, and Rama's ape got itself some lower teeth. Simons published this observation in 1964 in *Proceedings of the National Academy of Sciences*, and restated his conclusion that *Ramapithecus* almost certainly represented an early hominid. "This determination," he proclaimed, "increases tenfold the approximate time period during which human origins can now be traced with some confidence."[23] In a science in which practitioners have always vied to "push back the age of man's origin," this must take the prize for the biggest push of all time.

But it wasn't so much the age of the fossils that was so compelling as the very complete story they seemed to tell. "The evolutionary shift in a major adaptive zone indicated in the case of *Ramapithecus* by its reduced snout and anterior teeth (premolars, canines and incisors) may well correlate with an increased use of the hands and the incipient development of bipedality,"[24] said Simons and Pilbeam in a paper published in 1965. "Also, food must have been *prepared* for chewing by non-dental means," noted Pilbeam a year later.[25] "Hands were probably used extensively and perhaps tools as well," he added. *Ramapithecus* might have been fully bipedal and fully terrestrial, he speculated, which "would account for its widespread distribution." Pondering on why *Ramapithecus* fossils are so rarely found, Pilbeam extended this scenario and offered the following explanation: "Perhaps because they were creatures with a very low population density, perhaps because they were already hunters."[26] Pilbeams speculation is based on the "pyramid of nature," wherein hunters are at the apex and are few in number while the hunted are at the base and are plentiful.

Here, then, was a very complete picture of an animal—not just what it looked like, but also how it lived. And all based on a few fragments of upper and lower jaws and teeth. "Yes, what you see there is the complete Darwinian view of human origins, the complete package,"[27] Pilbeam now observes. "What we saw in the fossils was the small canines, and the rest followed, all linked together somehow. The Darwinian picture has a long tradition, and it was very powerful."

Darwin made only passing reference to human evolution in his 1859 classic *The Origin of Species*, but discoursed at length twelve years later in *The Descent of Man*. Without benefit of a fossil record to guide (or constrain) him, Darwin penned a description of how humans might have arisen from apes. He was concerned to explain man's upright posture, his manipulative skills (with tools and weapons), and his diminutive dentition—all of them factors in man's exalted status on this planet. "Man could not have attained his present dominant position in the world without the use of his hands . . . But the hands and arms could hardly have become perfect enough to have manufactured weapons, or to have hurled stones and spears with true aim, as long as they were habitually used for supporting the whole weight of the body . . . or so long as they were especially fitted for climbing trees."[28]

Man therefore became bipedal. But how was the change in dentition and size of jaw to be explained? "The early male forefathers of man were . . . probably furnished with great canine teeth; but as they gradually acquired the habit of using stones, clubs, or other weapons, for fighting with their enemies, or rivals, they would use their jaws and teeth less and less. In this case the jaws, together with the teeth, would become reduced in size." In other words, it is all tied together: small teeth imply the use of tools and weapons, which demands improved manipulative skills, which in turn implies an upright posture.

"All you needed in a fossil was small canines and the belief that it was a hominid, and everything else followed as a functional package,"[29] says Pilbeam. "This whole view reflects the expectation that the earliest hominid is already a pretty special creature, that it is already well on the way to being a human. It is very much a cultural animal."

Simons' view now of these Darwinian elements in the earlier papers with Pilbeam is unequivocal. "That's David. I never really thought *Ramapithecus* was a tool-maker. I didn't believe they had to be in order to be hominids. David was educated in an anthropology department. I wasn't; I learned in paleontology. I am not as interested in trying to re-create the behavior of an animal of which you only have fragments of jaws and teeth."[30] Simons is vehement about the distinctions between him and his coauthor: "I don't allow myself the luxury of speculating about an animal that doesn't any longer exist."

But any such differences between Simons and Pilbeam were in the 1960s very much in the background and little spoken about. Unanimity of view was both deeply felt and strongly expressed.

By 1968 yet another anatomical feature had been discovered that was thought to link *Ramapithecus* with *Australopithecus* and modern humans. All have a thick cap of enamel on their molar teeth, while chimpanzees and gorillas have thin enamel. It was naturally—but as it later turned out, wrongly—assumed that the African apes represented the primitive condition and that the "human line" had developed the specialization of thick enamel caps on the cheek teeth. No one knew at the time that orangutans also have thick-enameled cheek teeth. And no one took the trouble to see what pattern obtained in the rest of the Miocene apes: thick enamel is in fact a common feature, a primitive, not specialized, condition. But in 1968, the notion of *Ramapithecus, Australopithecus*, and modern humans sharing this "specialization" fitted nicely with the then-orthodox view of human origins and with the idea that apes are rather primitive creatures. In fact, Pilbeam wrote recently, "The presence of thick enamel on the molars of *Ramapithecus* . . . became the most compelling evidence linking this latter hominoid with thick-enamelled *Australopithecus*."[31]

The Yale duo's ideas on the timing of the divergence between apes and humans were developing too, principally as a result of the picture they saw emerging from the revision of the Miocene apes. Among these fossils Simons and Pilbeam thought they could recognize creatures that were directly ancestral to the modern great apes. "At this time—20 million years ago—three separate species of a genus called *Dryopithecus* were living which were most probably ancestral to the chimpanzee, the gorilla and the orang-utan,"[32] wrote Pilbeam in 1968. "I think that these ancient pongid species were already too specialized, already too committed to their ape-dom, to have produced hominids. Some day therefore we must expect to find hominids even earlier than *Ramapithecus*; it is even possible that the hominid and pongid lines have been separate for 30 million years and more." So, for a brief time at least, paleoanthropologists went around talking of an extremely ancient origin of the human line, and this view quickly found its way into textbook diagrams.

Louis Leakey liked the idea, of course, because, as he had already pointed out, if *Ramapithecus* lived 15 million years ago in Asia, with "*Kenyapithecus*" existing even earlier in Africa, then the first hominid of them all must be considerably more ancient.

But historically, the interesting issue here is the fact that Simons and Pilbeam were willing to see in 20-million-year-old fossils foreshadowings of modern apes, even though the fossil record was virtually void between the two points. This notion was very deep-seated in their approach, and was articulated in their revision of the Miocene apes, published in 1965: "It . . . seems unlikely that in such a biologically successful group as the higher primates very many taxa have ever become extinct except by time-successive transition into later species."[33] The effect of this assumption was simple: if you expect ancestral lines to run clean and uncluttered through vast tracts of time, you will expect to find ancestors very readily in the record if you look for them. "We found exactly what we expected to find,"[34] acknowledges Pilbeam.

Although, again, the revision paper was coauthored, Pilbeam claims that the influence here was mainly Simons', though he agreed with it at the time. "The scheme is very clearcut, with ancestors and descendants linked confidently,"[35] he now comments. "Few lineages become extinct, the hominoids being seen as a straightforwardly simple group." Pilbeam now describes this simple, ladderlike vision of evolution as quaint. "It is now clear that evolution is much more like a bush than a ladder. You simply cannot draw long lines through time as we were doing."[36]

The first decade of the resurrection of *Ramapithecus* had, by all accounts, been a great success story in the annals of paleoanthropology. These long-neglected fossils had been rescued from obscurity by the recognition that they had several humanlike features: the small canines and other anterior teeth, the curved dental arcade, the short face, and the thick enamel caps on the molar teeth. Modestly endowed thus, Rama's ape took on the mantle of hominid-dom, and all that this implied: it was a bipedal, cultural animal, skilled at manual manipulation and highly social. In other words, the fossils fitted the prevailing hypothesis for human origins, which was essentially Darwin's hypothesis. And most paleoanthropologists considered it a convincing story. With champions behind *Ramapithecus* of the academic stature of Simons and Pilbeam—both were Yale professors, one (Simons) with a doctorate from Oxford University and the other (Pilbeam) with a doctorate from Cambridge University—the story could hardly be otherwise.

From 1970 onward, however, many changes were afoot in paleoanthropology, and life for *Ramapithecus* became ever more uncomfortable. For instance, the governing hypothesis for human origins shifted, and ideas therefore began to change about what kinds of anatomical features could be considered important indicators of hominid status. And new interpretations of some Kenyan fossils showed that *Ramapithecus* must be much more primitive than had been supposed. Pilbeam and Simons managed to maintain their support of *Ramapithecus*, however, mainly by adjusting their lines of argument in concert with the shifting evidence. Eventually, however, Pilbeam's previously rock-solid confidence began to crumble, and before the decade was out Rama's ape would be just that—an ape—in his eyes. Much of the rest of the paleoanthropological establishment began to follow Pilbeam's example, slowly at first but later with enthusiasm. But for Simons, the change of heart would be a long time in the coming.

The first thing that happened during this decade of change was what sociologists of science call a paradigm shift: an established major hypothesis was replaced by a new one. Specifically, the Darwinian model, which emphasized tools and culture as the main driving force of human evolution, was overturned in favor of a dramatically different hypothesis. Clifford Jolly, a British researcher at New York University, proposed the new hypothesis in a now classic paper in 1970, titled simply "The Seed Eaters." The term "classic" is used here, as in most fields of science, to mean that the paper is almost certainly wrong in every detail, except one: its underlying philosophy.

Jolly proposed that the whole dental and facial structure in early hominids was the result not of a cultural package, but of the biomechanics required for chewing small, tough seeds. Instead of the gnashing action that chimpanzees, for instance, use to eat soft fruits, the hominids needed to grind these small objects, said Jolly, between millstonelike teeth. And the grinding action required the lower jaw to be able to move sideways with respect to the upper jaw, thus crushing objects between the upper and lower molars, which must be shaped like little millstones. Such a jaw action is impossible in the presence of large, projecting canines. Therefore the canines had to be reduced in size. The foreshortened face happens to generate the most efficient biomechanics for this kind of grinding action.

Jolly's scenario goes further. Manipulative skills, including the development of the characteristic opposable thumb of humans, are honed by the demands of picking up small objects from the ground. And an upright stance of the trunk—not bipedalism, but a "preadaptation" to it—arises from the most efficient body position in which to pursue this feeding strategy, which is squatting. Much of the inspiration for Jolly's hypothesis came from observing the behavior of gelada baboons in Ethiopia, whose feeding encompasses some elements of the seed-eating model.

"Jolly's paper was an extremely important development for paleoanthropology," Pilbeam now says, "not so much for the details of what it said, but because it approached human origins in an entirely different way. It shifted the focus away from culture and onto food and feeding behavior."[37] Here, for the first time, was an argument that those important hominid features—bipedalism, shortened face, and manipulative skills—had arisen in the complete absence of culture. It was a landmark change in ideas about human origins.

One problem with the idea of the cultural early hominid, Pilbeam observes, was that there was little room for any kind of intermediate form, any kind of precultural hominid. With Darwin's model the engine of culture is there from the beginning, and one is always dealing with some quaint kind of primitive human. As Pilbeam puts it, "The early ones always come to look like the late ones in these arguments." With Jolly's model it is possible to conceive of the evolution of a bipedal ape equipped with fine manipulative skills, which then might or might not develop culture.

How did *Ramapithecus* fare in the midst of this paradigm shift? As Milford Wolpoff, a paleoanthropologist at the University of Michigan, put it: "The focus shifted from the front to the back of the jaw, and *Ramapithecus* remained a hominid."[38] In other words, the diminutive canines had once provided the key for access to hominid-dom, but now the large, millstonelike molars had taken over that role. And just as *Ramapithecus* scored positive on the canines, so too did it with the molars: it still qualified as a hominid. The fact that the entire intellectual base for judging the putative hominid status of *Ramapithecus* had changed, and yet the conclusion had remained the same, apparently caused Pilbeam and Simons little concern—initially, at any rate. "We should have been aware how flimsy our original arguments had been," says Pilbeam. "And that should have made us more cautious. But it didn't"[39]

Ironically, the arrival and development of the seed-eating hypothesis provided something of a rescue mission for *Ramapithecus*, because as the 1970s wore on there arose an ever-sharper edge of criticism which began to cut away some of the more traditional props of its hominid status. A principal target was the shape of the jaw. Had the seed-eating hypothesis not apparently confirmed the hominid status of *Ramapithecus*, the evidence on the shape of the jaw would surely have had a more rapid and devastating impact on the establishment position.

Simons, remember, had reconstructed the two partial halves of the upper jaw of *Ramapithecus* and shown that "the arc formed by the teeth is curved as in man, rather than being parabolic, or U-shaped, as in apes."[40] He had been applying Le Gros Clark's criterion for distinguishing between apes and humans; but, as so often happens, the simply dichotomy—either like an arch or U-shaped—proved to be simplistic. "The problem is," Simons now observes, "that fossils rarely are what you expect them to look like. That means that the dichotomies that you draw from modern forms—the shape of the jaw in apes and humans in this case—simply don't work when you get ten, twenty, thirty million years back into the record. The fossils are likely to be an unpredictable mixture of known and unknown forms."[41] But this wisdom of hindsight was to be a while in the coming.

One of the first developments to indicate that something might be wrong with the interpretation of the jaw shape came in 1971 from a British student, Peter Andrews, working with Louis Leakey in Kenya. Andrews, who later was to play a leading role in unmasking the true identity of *Ramapithecus*, had been schooled in paleoanthropology by Leakey as a second career, his first being in forestry. A degree in anthropology from Cambridge University gave him his intellectual background, and being at Leakey's elbow provided the hands-on experience with original fossils—when he could get access to them, that is. "Louis used to keep the fossils locked in a safe in his office at the museum, and he had the key,"[42] recalls Andrews. "It was very difficult to get to see the specimens. You had to catch him, and you had to catch him in a good mood, and he had to let you into the safe."

It was on just such an occasion late in 1970 that Andrews grasped the opportunity to examine the famous *Kenyapithecus wickeri* upper jaw and a lower jaw found in the same locality. "Both Leakey and Simons had separately identified this lower jaw as that of a species of *Dryopithecus*—a Miocene ape. I was looking at these specimens, put the lower jaw with the *Kenyapithecus* upper jaw, and saw that they fit perfectly—the overall shape, the detailed anatomy, everything. They might well have belonged to the same individual." This observation led to the publication of a small paper in *Nature* the following May, which simply stated that *Kenyapithecus wickeri* now had a lower jaw.

But the implication was obvious. Here was a lower jaw that had been identified as that of an ape because of its primitive structure, but in fact it belonged to a putative hominid. Ergo, the hominid must be a good deal more primitive-looking, a good deal more apelike than had previously been supposed. The discovery didn't particularly jolt Pilbeam's confidence. It meant that "we had a very primitive hominid. Theoretically there would have been no problem with that," he now says. "I didn't spend much time talking about it with Elwyn."[43] Andrews remembers that Simons was less than pleased with the *Nature* paper, though he did not dispute the interpretation.

The most important development to stem from Andrews' observation, however, was a collaboration with Alan Walker in which the two made a reconstruction of the *Kenyapithecus* jaws. Walker, who later was to become Richard Leakey's closest colleague, is, like Simons, an accomplished sculptor and claims to have a keen sense of three-dimensional shapes. Working with plaster replicas of the originals, some made in mirror images, Walker and Andrews produced a jaw shape quite at variance with that accepted for *Ramapithecus*. The animal had "nearly straight tooth rows, rather than widely separate and curved tooth rows as are found in modern man,"[44] they reported in the 3 August 1973 issue of *Nature*. It was clear, they continued "that it did not have the rounded dental arcade postulated in previous reconstructions." This observation was to trigger a dispute with Simons that continues to this day.

Walker and Andrews did not actually say that *Ramapithecus* could therefore not be a hominid, but what they were describing was a more or less classic Le Gros Clark–ian shape of an ape's jaw. "They were trying to put the wind up Simons, I guess,"[45] Simons now says. "But they made the silliest mandibular reconstruction ever produced. It has six different errors of orientation in it. They didn't take adequate account of how crushed and distorted it was. Alan's talents as a sculptor sure didn't come out in that reconstruction." Simons wrote a very long letter to Walker and Andrews pointing out where he considered they had gone wrong. He also prepared a lengthy manuscript criticizing the reconstruction, but it was never published because, he says, it was so destructive. Pilbeam's reaction was different. "I thought what they had done was more or less right. I thought the best response would be to say, That's correct. But I went along with the general argument. I felt uneasy, but not sufficiently to break ranks."[46] By this time, unanimity between Pilbeam and Simons was less strongly felt than it once was, but was nevertheless still publicly expressed.

Other researchers began to criticize Simons' original *Ramapithecus* upper-jaw reconstruction, primarily on the ground that its fragmentary nature could not permit a confident interpretation of the shape. Among these critics were Milford Wolpoff and two of his students, David Frayer and Leonard Greenfield. "We pretty much ignored them," says Pilbeam.

Nevertheless, a degree of uncertainty was beginning to stir into Pilbeam's mind. The shift in the governing human-origins hypothesis began to be reflected in his writing: for instance, he said, *Ramapithecus* "may well not have been a biped," and "it seems not to have been a habitual tool-maker."[47] And he was also less dogmatic in saying for certain that *Ramapithecus* was a hominid: "I think there is at least a 75% chance that *Ramapithecus* is ancestral to later *Hominidae*," he wrote in 1972. He also changed his mind about the antiquity of the first hominid: "Other workers argue for a separation of ape and human lineages still earlier than fifteen million years ago. I have done so myself, although I now think that any divergence much older than fifteen million is quite improbable," he said in the same paper.

Through the mid 1970s, Pilbeam began to realize that the fossil material then available simply wasn't adequate to support the kinds of sweeping conclusions that had been made. His uncertainty grew to a point at which he more or less refrained from saying anything at all about *Ramapithecus* and its putative hominid status. "His colleagues began to make cracks about it,"[48] remembers Simons. But Pilbeam remained silent, needing something to push his thoughts in a new direction.

That something—a new fossil—came to hand in January 1976 during Pilbeam's field expedition to the Potwar Plateau of Pakistan, which is about 400 km from where Lewis found the original *Ramapithecus* upper jaw. The sediments there are a mile thick, the massive accumulation of debris washed down from the Himalayas over millions of years, and they have yielded a rich collection of Miocene hominoids. On that January day, Wendy Barry found the complete left side of a *Ramapithecus* jaw, the best one the expedition had recovered to date. But more important, it clearly matched a fragment that Martin Pickford, another expedition member, had found on the last day of the previous year's field season. Very soon several more fragments of the same jaw turned up, giving the most complete *Ramapithecus* specimen ever found. "Many things ran through my mind," Pilbeam wrote later, "feelings of exhilaration and happiness for the expedition, for the discoverers, satisfaction for myself; but more than anything my mind was beginning to think along new lines, because I could see that our previous belief about *Ramapithecus* and the whole story of human origins needed rethinking."[49]

What Pilbeam could actually see was that the little jaw he held in his hand had a kind of truncated V-shape, not the rounded arcade that it "should" have had. "I think I had always been unhappy about the curved arcade of the original reconstruction," he now says, "because you can't see the midline. It turned out that Len Greenfield and the others had been right, but I needed to work it out for myself."[50]

The principal impact of that little *Ramapithecus* jaw was that it dispelled the notion that the Miocene hominoids were either apelike or humanlike. "This was the first mandible we could look at and be sure of the shape, because of its completeness," he explains. "It was neither like an ape nor like a human. It was something quite different." Pilbeam now felt he had something concrete to work with, and it was the beginning of the break with the established view on *Ramapithecus*.

Simons, meanwhile, stayed fast, more or less. In an article written for *Scientific American* not long after the discovery of Pilbeam's new jaw he stated that the pathway of human origins can be described over the past 14 million years "with little fear of contradiction."[51] He said that he first noticed the V-shaped jaw of Miocene apes in 1967, and that the *Ramapithecus* jaw is a good intermediate between that and the rounded arcade of humans. And he reaffirmed his conviction that *Ramapithecus* is "a very early hominid." Pilbeam now describes this article as unfortunate. "It was not a good idea intellectually to so forcefully repeat an argument that, in my opinion, was weak. He was locking himself in with it."[52]

Simons left Yale in April 1977, to become director of Duke University's Primate Center in North Carolina. Communication between the two former colleagues diminished dramatically, and the gap between their intellectual positions widened even further. Pilbeam was now free to break completely and publicly with his and Simons' former position. But it wasn't just the new fossils that forced him to do it. He found he could no longer ignore the increasing weight of biochemical evidence, which implied that he and Simons had been badly wrong since the beginning. In the end, this new and "heretical" line of evidence would be crucial to Pilbeam in particular and to paleoanthropology as a whole.

CHAPTER 6

RAMA'S APE: DESTROYED

The decade leading up to Elwyn Simons' departure from Yale and David Pilbeam's defection had witnessed the growth of an entirely independent line of criticism of the putative hominid status of *Ramapithecus*. This came from biochemists and molecular biologists who insisted that inferring relationships from fossils was fraught with potential error and that a much more reliable method was to compare proteins and nucleic acids from living species: in this case, from humans and African great apes. These molecular methods indicated a separation time between apes and humans much more recent than the 15 to 30 million years implied by the fossils; specifically, more like 5 million years.

The discrepancy was so great and the confidence in the predictive power of the molecules so solid that it encouraged Vincent Sarich, a leading proponent of the technique, to proclaim in 1971 that "One no longer has the option of considering a fossil specimen older than about eight million years a hominid *no matter what it looks like.*"[1] In other words, he did not care whether *Ramapithecus* looked like *Australopithecus* or even *Homo sapiens*. It was simply too old to be a hominid. Period. A statement more calculated to raise the blood pressure of paleoanthropologists could hardly be imagined.

The response of Simons, Pilbeam, and their colleagues to the biochemical evidence was first to ignore it, then to ridicule it. Eventually Pilbeam came to accept the evidence as the most important guide to events in human prehistory, more important even than the fossils. Simons was unimpressed, and remains so to this day.

The story of the intrusion of molecular evidence into paleoanthropology has three beginnings, two of which were in a sense false starts. The first was at the outset of the century, when George Henry Falkner Nuttall, a professor of biology at Cambridge University, speculated that the chemistry of blood proteins might be used to determine the genetic relatedness between animals, specifically the higher primates, including man. The idea, based on the work of the great Paul Ehrlich, whom he met while doing his doctoral research in Germany, was that as distance of genetic relatedness increased, so too would the difference in the chemistry of the blood proteins. "The persistence of the chemical blood relationships between the various groups of animals serves to carry us back into geological times, and I believe . . . that it will lead to valuable results in the study of various problems of evolution," he wrote in the *British Medical Journal* in 1902.

Nuttall did some preliminary experiments in which he showed that humans are more akin to the higher primates of the Old World than to those of the New World. In so doing, Nuttall was simply reaffirming what both Charles Darwin and Thomas Henry Huxley had inferred forty years earlier from anatomical comparisons. Darwin and Huxley had in fact gone even further, and suggested that in spite of superficial appearances, detailed anatomical comparisons revealed the gorilla and chimpanzee to be closely allied to humans, with the third of the great apes, the Asian orangutan, more distantly related to humans.

Nothing much happened for the next sixty years, except perhaps that people tended to forget the genetic intimacy between humans and the African apes. There was for a while an implicit assumption —often expressed in diagrams of the hominoid tree, even if not in words—that the three great apes were closest together genetically as well as in superficial appearance and behavior.

So when at the start of the 1960s Morris Goodman, of Wayne State University in Detroit, began producing hominoid evolutionary trees showing again the human/African-ape affinity, essentially using Nuttall's approach on blood-serum proteins, paleoanthropologists were more surprised than they should have been. Goodman, being a logical scientist, therefore decided that it was folly to classify, as zoologists have always done, all the great apes under one family, the Pongidae, and humans in splendid isolation in their own family, the Hominidae. Clearly, because of their genetic propinquity, humans, chimpanzees, and gorillas belong in the same family, thought Goodman, and he said so at a 1962 Burg-Wartenstein symposium in Austria.

"The professional taxonomists blew their top at this suggestion," remembers Sherwood Washburn, a leading American paleoanthropologist who organized the meeting. "If only he had left the names alone I think people would have been more sympathetic to the whole biochemical approach. The people who come in from biochemistry always want to change the names."[2]

Washburn, who was a student of Earnest Hooton's at Harvard, has long argued like Huxley that humans and African apes are closely related, principally on the basis of detailed anatomical comparisons. This conclusion led him to believe that, contrary to what people like Simons and Pilbeam were saying, the divergence between humans and apes was in fact a relatively recent event. He said as much at the 1962 meeting: "Most of the characters of *Homo* seem to have evolved well within the Pleistocene, and there is no need to postulate an early separation of man and ape."[3]

The limited fossil record of the time offered no obvious support for this hypothesis, but Washburn immediately recognized the biochemical approach as a new and different source of evidence that might help. Within a couple of years of the Burg-Wartenstein meeting, Washburn, who was a professor at Berkeley, was encouraging a graduate student to see if protein chemistry might provide the answers he was seeking. That student was Vincent Sarich, a chemist who decided to pursue anthropology instead. Washburn wanted numbers—that is, dates on the branches of the tree—which Goodman had not provided.

Sarich very soon discovered that a great deal of the most relevant data were already in the scientific literature. Much of this was Goodman's data, of course. But the basic ideas for using data from differences in protein chemistry for constructing evolutionary trees came from Emil Zuckerkandl and Linus Pauling on one hand and Walter Fitch and Emanuel Margoliash on the other. What remained was a demonstration that the concept worked; that proteins could be used as a clock to mark the times at which each branch diverged from the stem of a family tree, the hominoid family tree in particular. Sarich joined forces with Allan Wilson, a young biochemist on the Berkeley faculty, and encouraging results were not long in coming.

At this point, early in 1966, Washburn wrote to Simons telling him he had a secret weapon with which he was going to prove a recent ape/human divergence. Did Simons wish to join him in a controversy? he asked. The battle lines were drawn.

All these initial approaches to molecular phylogeny, as the drawing of family trees from biochemical data is called, depend on one assumption: that once an ancestral stock splits to form two separate species, their proteins steadily and regularly accumulate differences through mutation, and the longer the time of separation, the greater will be the difference in protein structure. When Goodman examined the protein chemistry of the hominoids—which includes the gibbons as well as the great apes and humans—he was able to get a general shape of the tree. It showed the gibbons branching off first, the orangutan next and the gorilla, chimpanzee, and human being all rather close together. But he noticed that the overall differences in protein chemistry were rather small, and certainly smaller than would have been predicted from the separation dates deduced from the fossil record. He therefore concluded in 1963 that the accumulation of differences in protein structure was not necessarily steady, and in this case had demonstrably slowed down.

When in 1966 Sarich and Wilson embarked on their joint project, they had Goodman's conclusions before them in the literature. They therefore set themselves clear and limited goals: "We wanted to know whether the marked similarities among hominoid albumins [blood proteins] were due to a slowdown in their rates of change, and if not, what sort of time scale for ape and human evolution could be extracted from the data."[4] Within a year they had answers: there had been no slowdown in the rates of change, they concluded, and so it was indeed possible to set up a time scale of hominoid evolution. Sarich and Wilson hoped to publish their results in three separate papers as a logical sequence.

The first paper simply described the basis of the technique as applied to hominoids, which the journal *Science* published with alacrity in its 23 December 1966 issue. The third paper, which *Science* also readily accepted and published on 1 December 1967, presented Sarich and Wilson's inferred time scale: " . . . man and

African apes shared a common ancestor 5 million years ago, that is, in the Pliocene era."[5] But the middle paper of the sequence, which, says Sarich, contained "the only really original and critically necessary material,"[6] was rejected by *Science*. The reviewers of the paper considered that it said nothing new or important.

What the paper in fact presented was the "rate test," which is a way of examining data on differences in protein chemistry from several related species and determining whether the rate of change of the proteins has been regular or irregular. In this case it demonstrated that the rate of change in serum albumins in hominoids is steady; that it is clocklike, and had not slowed down, as Goodman had suggested. No longer was the clock resting on an assumption. It was clearly supported by demonstrated fact. Disgusted with their treatment by *Science*, Sarich and Wilson asked Washburn if he would "communicate" their paper to the *Proceedings of the National Academy of Sciences*, where it would be published quickly and without review. Washburn obliged, and the paper appeared in the fall of 1967, to remain unread by most of the paleoanthropological community.

"One wonders if the continual accusation that we have somehow 'assumed' the clock and imposed it upon the data stems in large part from this original misperception on the part of the reviewers *Science* used," Sarich now muses. "Ignoring that article makes it easier to assume that we assumed the clock—no doubt of that." It is true that most of the critics over the next decade would frequently charge that Sarich and Wilson "assumed" constancy of change as the basis of the clock. But it is probably also true that the negative reaction to the message of the clock was so strong that nothing at all would have eased its eventual acceptance.

Armed with their results, Sarich and Wilson tried to persuade Goodman that the close hominoid relationships indicated by his protein results were in fact real and he should not allow his interpretations to be influenced by what the paleoanthropologists were saying on the basis of the fossil record. Goodman, however, stayed fast with the idea of a slowdown in the molecular clock, and continues to do so to this day. For instance, at a symposium in Toronto in January 1981 he said firmly: "Humans and chimpanzees show not only an especially close kinship but also a markedly slow rate of protein evolution."[7] Sarich and Wilson were dismayed at

Goodman's intransigence. "I'm pretty intolerant of the idea of a slowdown,"[8] says Wilson. "Why does he say it? He says it because he is afraid to take on the paleoanthropologists." "He conceded to the paleontologists their coveted arbitrators' mantle,"[9] adds Sarich. "[They think that] if we cannot see it in the fossil record, it never really happened."

In the decade and a half that followed Sarich and Wilson's first papers, the biochemical approach encompassed new and more powerful techniques, some of which involved analyzing the DNA structure itself. It was like turning up the magnifying power on a microscope 1000 times. And yet the story did not change very much, as Sarich noted with some satisfaction in 1982: "Thus, what was a tentative best guess in 1967 had developed into a virtual certainty in 1970—and remains so to this day." Wilson's assessment at this point was the following: "it is reasonable to argue, as we have done, for a late separation of human and African-ape ancestral lineages. One should not be dogmatic, however, about the *exact* time of separation."[10] For Sarich and Wilson, that separation time has always hovered around 4 to 6 million years ago.

Paleoanthropologists' initial reaction to the messages in Sarich and Wilson's 1967 paper was mixed. "Most people simply ignored it,"[11] recalls Sarich. "But a small number vilified it. What was published was, however, a greatly sanitized version of what was said in private."

Pilbeam's first public comment about the molecular evidence came in a paper published in the journal *Nature* in 1968, in which, incidentally, he also took the opportunity to needle Louis Leakey. "Recently several authors have stated their view that the African apes shared a common ancestry with Hominidae as late as 5 million years ago," he wrote. "If this theory is correct, *Ramapithecus* cannot be a hominid, as Leakey, Simons and I believe. None of the *Dryopithecus* species can be ancestral to any living pongids . . . I am inclined to accept, however, the fossil evidence for the moment."[12] It was a measured statement, fairly typical of Pilbeam's approach.

Simons was a little more pointed. Also in 1968 he wrote: "If the immunological dates of divergence devised by Sarich are correct, then paleontologists have not yet found a single fossil related to the ancestry of any living primate . . . I find this impossible to believe. It is not presently acceptable that *Australopithecus* sprang full-blown five million years ago, as Minerva did from Jupiter, from the head of a chimpanzee or gorilla."[13]

In the same paper, Simons puts Sarich and Wilson in their place —that is, as outsiders to the game. "Students of human origins will know, however, that the story of hominid origins begins much earlier than this, since hominids of the genus *Ramapithecus* date back to the late Miocene, about 14 million years ago."

Even Louis Leakey joined in the fray, admitting first that "I am not qualified to discuss the biochemical evidence,"[14] and then going on to assert that it must be wrong because it was at variance with the fossil record. "The fossil record shows clearly that . . . between 12 and 14 million years ago, there were already present: a) a true member of the Family Hominidae, *Kenyapithecus wickeri*, b) a true member of the Family Pongidae, represented by both the genera *Dryopithecus* and *Proconsul*, c) true members of the family Hylabatidae [gibbons], represented by a primate allied to *Limnopithecus*, and *Propliopithecus* . . . [etc. etc.]," asserts Leakey in a plethora of paleontological nomenclature. "The date of separation suggested by Wilson and Sarich, i.e., only five million years ago, is not in accord with the facts available today." It is clear what, in Leakey's eyes, constitutes facts and what does not.

This initial line of criticism by the paleoanthropologists is unequivocal: the biochemistry is wrong because it doesn't agree with the fossils. Period.

A second form of attack then began to develop, with the idea that, as Milford Wolpoff once so clearly stated it, "the 'clock' *should not* work."[15] Note, incidentally, the use of quotation marks around "clock." "Yes, people always did that,"[16] remarks Sarich. "It meant, Here, laugh at this; you don't have to take it seriously."

John Buettner-Janusch, then of Duke University, began the argument in a lecture to the Division of Anthropology at a New York Academy of Sciences meeting in October 1968. "I tell my students that there is a vast amount of garbage floating on the stream of physical anthropology,"[17] he said, leaving little doubt of his viewpoint. He then described in outline the general approach to molecular phylogeny. "An exercise of this sort, which I deplore, is based on a number of rather simple (indeed simple-minded) assumptions . . . We must assume that the mutations . . . occurred at a uniform rate or, at the very least, randomly since phyletic divergence of the two lineages. . . . We are forced to ignore certain more realistic assumptions."

Buettner-Janusch then went on to remind his audience that if Sarich and Wilson had taken time to look at the fossil record they would have seen that they must be wrong. And he finished, to general acclaim, with the following: "I object to careless and thoughtless statements about evolutionary processes in some of the conclusions drawn from the immunological data mentioned. . . . Unfortunately there is a growing tendency, which I would like to suppress if possible, to view the molecular approach to primate evolutionary studies as a kind of instant phylogeny. No hard work, no tough intellectual arguments. No fuss, no muss, no dishpan hands. Just throw some proteins into the laboratory apparatus, shake them up, and bingo!—we have an answer to questions that have puzzled us for at least three generations."

The appeal of this second line of objections to molecular phylogeny is simple and direct. "The clock does not keep good time," as Simons forcefully put it to a gathering of paleoanthropologists in Nice, France, in 1976. To judge from the tumultuous applause he received, most there agreed with him. And they seemed to have good cause.

There is in fact no obvious reason why the accumulation of mutations in protein molecules should always be regular through time, no reason why the molecular clock should tick metronomically. Biologists have long observed that evolution is a rather irregular process, with modification of form and function occurring in an unpredictable manner, depending on changes in the environment, for instance. There is nothing uniform or inexorable about natural selection. The structure of protein molecules is equally subject to natural selection, and therefore might undergo substantial modification at some points in history and little change at others. But proteins can also accumulate mutations that don't immediately affect function. These occur at a steady rate and are known as neutral mutations. The buildup of neutral mutations can therefore be seen as a basis for a protein molecular clock. How regular the clock is will depend on what other changes are taking place at any particular time.

"The molecular clock is antithetical to a hundred years of studies on evolution,"[18] observes Sarich. "No one thought that anything in evolutionary biology could proceed in a regular, clocklike manner. Slow yes. Gradual, yes. But no one considered the process was clocklike." His position on this is clear: you must always test whether your protein or proteins operate like a clock. "Starting a priori, you would not necessarily expect there to be a clock. Yet in some cases it demonstrably exists," he says. "The clock works where it works, and not where it doesn't. Simple. You have to demonstrate that it works, and then you're OK." Which is one reason he and Wilson became more than a little irritated when critics continually asserted, as did Buettner-Janusch, that regularity of change had been assumed. "Everyone 'knew' that we had *assumed* constancy in the clock," quips Sarich. "We know that can't be right, they said. Therefore Wilson and Sarich are idiots." "I know that when I talked with people like Alan Walker and Owen Lovejoy about the clock they didn't offer any reason why it couldn't be true,"[19] says Wilson. "They just felt it was somehow obvious it wasn't true."

An interesting example of this attitude is contained in an unusual footnote to a paper by Sarich in a conference volume on Old World monkeys, published by Academic Press in 1970. Sarich's paper, entitled "Molecular Data in Systematics," was one of three papers on phylogeny in the first part of the volume. At the bottom of the first page of his paper was the note: *"Dr. Sarich's contribution does not necessarily reflect the views of the participants of this Conference, unlike the other two chapters of part I. Eds."* "Yes," Sarich now jokes, "it was like a government health warning, wasn't it?"[20]

Now, it is all very well for Sarich and Wilson to say that the paleoanthropologists should have accepted the clocklike nature of the biochemical methods. But no scientist likes to rely on a method that has the aura of a black box about it, especially if it gives unwelcome answers. And there were genuine uncertainties about the approach at the time. Given the existence of the uncertainties in the late 1960s and '70s, it was sufficient for any critic of the technique simply to raise the issue of uncertainty, whereupon he would step back, apparently secure in the knowledge that he had thereby demonstrated that it had no use at all. "There was the feeling engendered that because some uncertainty exists (it is, one needs to be reminded, a probabilistic, not metronomic, clock) we need not take the entire approach seriously,"[21] says Sarich.

Pilbeam engaged in this tactic in the early 1970s, when he was still a firm supporter of *Ramapithecus*. In a long paper in the December 1971 issue of the journal *Evolution*, Pilbeam and a Yale biochemist, Thomas Uzzell, pointed out that the accumulation of mutations could not be clocklike because, as time moved on, an increasing number of events would be hidden (two mutations at the same spot would, for instance, show up as only one). They concluded that therefore "unless uniform rates of evolution can be demonstrated, or unless they have been assumed, biochemical data do not provide an adequate basis for discarding dates of divergence based on fossil data."[22]

"They were absolutely correct to point out this element of uncertainty,"[23] Sarich now says. "But if you correct for it you get even shorter divergence times, not longer. I wrote them a long letter about it before their manuscript was published, but they didn't take any notice of it. Essentially, their article says, 'Where there's smoke, there's fire.' "

As so often happens with disputes in academia, there developed a clear polarization of positions: it became the Berkeley school versus the Yale school. "By reducing it to schools of thought," Adrienne Zihlman told the annual meeting of the Southwestern Anthropological Association in April 1982, "everyone was assured they would not have to learn biochemistry after all. They would not have to take Sarich and molecules seriously and could ignore the abundance of information burgeoning from many laboratories and reduce it to one person's fantasy about albumin molecules."[24] She remembers that "If you were at Berkeley, as I was, elsewhere you were treated like a Moonie."

Simons insists that the doubts were real, that it wasn't a blind polarization spurred by ignorance. "You have to remember that we were well acquainted with what the biochemical evidence was,"[25] he now says. "It was accompanied by a chorus of papers from biochemists and mathematicians who were saying that the rate of biochemical change wasn't a straight line. It didn't look to us like a methodology that was gaining much credence." There was nothing new in the technique, he points out, citing Nuttall's work at the beginning of the century. "And I knew about these methodologies when I was fourteen, because I did a junior high school project on immunological-distance measures between mammals. Sarich and Wilson don't know that. So it was never news to me, this method. The only thing that was new was the vociferousness and the rock-hard allegation and charge that these dates were absolutely solid and correct. That was what was new."

The tenor of debate was indeed sharp, with each side accusing the other of profound, incurable ignorance. But personality played its part too. Simons, for instance, is not known for his timidity. And Sarich, by his own admission, is less than diplomatic. "My diplomacy consists of keeping my mouth shut,"[26] he says, "and that is difficult enough." Simons suggests that "part of the reason the thing was somewhat acrid at the beginning was because of Sarich's aggressive and cocky personality."[27]

Huge in stature, voice and opinion, Sarich has infuriated many a paleoanthropologist by his unrestrained mode of discourse. His most famous outburst—the "one no longer has the option . . ." statement mentioned earlier—brought dismay from all quarters, even in Berkeley. Pilbeam and a Yale graduate student, Glenn Conroy, described it coldly as being "best interpreted as overenthusiasm for a new technique." Simons called it "outrageous." And even Washburn admits that "It was the dumbest thing Vince ever said."[29]

But Washburn also says that had Sarich been a more timid personality, he would surely have succumbed to the barrage of criticism, and accomplished less. "Vince is a very strong person, which in the circumstances was very lucky. He wanted to convert people faster than I think it was reasonable that they should be converted." Washburn once tried to persuade Sarich to write an article that would explain step by step the molecular-clock idea, the rate test, and so on. "You still haven't given us a paper that the average anthropologist could read and see how you get these conclusions," he told Sarich. "No, Sherry. That's all published. People should read what's published, and they should accept it" was Sarich's reply.

Paleoanthropologists didn't accept it, of course; not right away. But in spite of the very negative reception the molecular data were given, they began to have an impact, even if paleoanthropologists have generally been less than open in admitting it. Specifically, the dates of 30 to 40 million years for hominid origins that were common in the literature through the late 1960s and early 1970s simply vanished, to be replaced with 14 or 15 million years. "Yes, we trimmed our dates, to be as respectable as they could be,"[30] recalls Pilbeam. "People do, don't they, in these kinds of circumstances, just in case there was something in it. You trim off the bits that are not so easily defensible. We said to ourselves, OK, we are convinced *Ramapithecus* is a hominid; it goes back to fourteen or fifteen million years; so we'll stick with that."

Simons more or less agrees. "It seemed to me that the split point was later, more like fifteen million than twenty or so, just because the biochemical evidence must mean something,"[31] he now says. "So we moved the date, but no more than we thought we warranted."

With this change came an instant rewriting of history in many people's minds. "Pilbeam and Simons often expressed astonishment that we would say that they ever thought the human/ape divergence was as old as thirty million years ago or more,"[32] says Wilson. "You can see it in Pilbeam's book—he has the Pongidae separating back in the Oligocene, around thirty million years ago. And yet they acted as if they had always said it was fifteen million. They functioned as if we didn't exist. They just ignored us."

So, although the paleoanthropologists were prepared to trim their dates in the face of the molecular evidence, they were not willing to go to the logical conclusion and concede that *Ramapithecus* was not a hominid. For that radical shift they demanded more fossil evidence, which came early in the 1980s.

Ironically, it was not the discovery of more *Ramapithecus* fossils that finally brought about the downfall of Rama's ape. Instead, it was another, closely related fossil ape, known as *Sivapithecus*. This extinct ape, which is a somewhat larger version of *Ramapithecus*, has often been recovered from the same geological deposits as its more famous cousin, in both Europe and Asia, and possibly in Africa too. Siva, by the way, is the Hindu god of destruction. In this case, Prince Rama was to be its victim.

Briefly, *Sivapithecus* undermined the putative hominid status of *Ramapithecus* in the following manner. The discovery and subsequent description of parts of two *Sivapithecus* faces—one in Turkey, reported in 1980, and the second in Pakistan, reported in 1982—allowed paleoanthropologists to recognize that this extinct ape is related to the living orangutan. Now, as *Ramapithecus* is clearly closely related to *Sivapithecus*, it too must be related to the orangutan. Ergo, being more closely related to the orangutan than to the African great apes, *Ramapithecus* is disqualified from being a hominid, because humans are more closely tied to the chimpanzee and the gorilla than they are to the orangutan.

The *Sivapithecus* partial face from Turkey had been discovered in 1967, but it wasn't properly analyzed until Peter Andrews turned to it late in 1976. He had just returned from a lively paleo-anthropological meeting in Nice, France, where, as mentioned earlier, Simons had been stoutly defending the hominid status of *Ramapithecus*. But also at the meeting there was an undercurrent of concern that one of the main props of Rama's ape—the thick enamel caps on the cheek teeth—might no longer be as secure as once thought. The Sinap face, named after the locality where it was found, might offer some clues, thought Andrews. It did.

A detailed study of the Sinap face, done in conjunction with I. Tekkaya, of the Paleontological Service, Ankara, showed that *Sivapithecus* was quite similar in many ways to the Asian great ape. "In the description of the maxilla [upper jaw] and dentition of *Sivapithecus meteai*, the closest comparisons in most cases were with the orang-utan,"[33] Andrews and Tekkaya were eventually to report. This paper came in for damning criticism from Elwyn Simons, who said that the specimen was crushed and distorted and the reconstruction unrealistic.

If Andrews' interpretation was correct, however, it would constitute a major breakthrough in the search for human origins, and one might have expected him to be eager to rush it into print. Instead, he waited for more than a year. "The problem was that in the very first paper I ever wrote, in 1970, I claimed to recognize oranglike features in the palate of another fossil ape, *Proconsul*,"[34] he now explains. "It was a mistake. I had been completely wrong, and so I was very cautious about 'finding' another orangutan." Eventually he did publish his findings on the Sinap face—in February 1980—saying that it had affinities with the orangutan, pointing out the close association between *Sivapithecus* and *Ramapithecus*, but saying nothing at all about the implications for the hominoid evolutionary tree.

In order to be really certain about the relationships within the tree, Andrews felt he needed to become completely familiar with the facial anatomy of the living great apes—a painstaking task of anatomical analysis that consumed two full years. During this period he met Jack Cronin, a former colleague of Sarich's, at a Primate Society conference in Bangalore, India. Being of the molecular-clock school at Berkeley, Cronin was excited about a possible *Sivapithecus*/orangutan association, because he recognized that it implied a human/ape divergence date more like what

Sarich and his colleagues had been promoting than the one embedded in the paleoanthropologists' conventional wisdom. The two men had a lot of opportunity to discuss their separate approaches, and one result was that Andrews for the first time began to take the molecular evidence seriously. A joint paper was planned, which for the first time would bring the fossil and molecular evidence together in harmony. The paper would say that *Ramapithecus* was not a hominid and that the ape/human divergence was about 5 million years ago. They would submit it to *Nature*, they decided.

A year passed, with both Andrews and Cronin distracted with their separate activities, the paper languishing incomplete. What spurred them to return to it was the news that Pilbeam and his team had discovered a *Sivapithecus* face in Pakistan, which, by all accounts, showed similar features to the one from Turkey. The manuscript was quickly finished, and finally submitted to *Nature* in November 1981. A month later, Pilbeam was passing through England and took time, as he usually did, to visit Andrews at the Natural History Museum, in London. "I showed David our manuscript," says Andrews, "and he was amazed and aghast. It turned out that he and Steve Ward had been working on similar lines, only in considerably more depth." Pilbeam already had a manuscript in press with *Nature*, which had been submitted two months earlier, but it was a short report on the Pakistan *Sivapithecus* face and did not go into the detail of Andrews and Cronin's paper. He and Ward had planned to prepare a more detailed paper later.

As so often happens in science, two independent research groups were simultaneously converging on the same conclusions. In this case, Pilbeam acknowledges, Andrews got there first.

Pilbeam's short *Nature* paper appeared in the 21 January 1982 issue. It was a brief, straightforward description of the face, which is code-named GSP 15000 and simply concluded that "there are several similarities between GSP 15000 and [the orangutan] that may turn out to be shared derived features."[35] Pilbeam hinted that work on the new face had led him to conclude that some of the anatomical features which had linked *Ramapithecus* with the known hominids *Australopithecus*—such as thick enamel, large molar teeth, and robust lower jawbone—might not actually be diagnostic of hominid status after all. This conclusion, he noted simply, "would have important consequences for the interpretation of hominid origins."

A more cautious offering could hardly be imagined, which caused some of his colleagues to wonder if Pilbeam had lost the ability to be certain about anything at all. "Well, having been wrong about relationships before, it is hardly surprising that I was cautious,"[36] he now says.

Andrews was invited by the editors of *Nature* to submit an editorial comment on Pilbeam's paper, and he readily acquiesced. "I was concerned that David had not gone far enough in his paper,"[37] remembers Andrews. So he articulated what had clearly been in Pilbeam's mind: "It thus appears that *Sivapithecus* (including '*Ramapithecus*') is part of the orang-utan clade. . . . In other words '*Ramapithecus*' can no longer be considered as part of the human lineage."[38] It was a landmark statement.

Meanwhile, Andrews' own joint manuscript with Cronin was held up in the editorial process at *Nature*. "A reviewer in America had blasted the manuscript, saying that there was nothing new in it, that the Sinap face was badly reconstructed and the interpretation incorrect,"[39] says Andrews. "It was a very sweeping criticism, and the manuscript was rejected as a result." Andrews decided that it was worth trying to persuade the editors that the reviewer was perhaps not as unbiased as he might be, and he explained once again why he considered the paper to be new and important. "To my surprise, they changed their mind, and published the paper as a review article in the 17th June [1982] issue." As planned, the Andrews/Cronin paper presented both fossil and molecular evidence, showing that they were now concordant.

The conclusion was inescapable: the molecular biologists had been right all along. It was there in black and white in *Nature*. By being associated with Cronin on this important paper, Andrews was being absolutely explicit about his confidence in the power of molecular data. Although he was principally a fossil researcher, he had not nailed his colors to the mast of the good ship *Ramapithecus*, and so he could abandon it relatively easily and openly acknowledge the utility of this once-heretical branch of science. For Pilbeam, however, it was different. "Yes, the molecular evidence was important in my thinking," he now says. "I could see there might be problems with it, but even by the late 1970s I could also see that we couldn't ignore it any longer. I knew it must mean something. And in the end I came to recognize that for some things at least, it was more reliable than fossils."[40] But, to the chagrin of Sarich, Wilson and their associates, it was to be a long time— about six years—before Pilbeam would acknowledge as much in writing. "Well, you wouldn't expect me to, would you?" says Pilbeam.

Simons, meanwhile, had been following events with some interest. In April 1980 he joined in a small gathering of researchers at the Duncan Hotel in New Haven, which Pilbeam had organized for discussions of the Miocene hominoid problem. Just a couple of months earlier the new *Sivapithecus* face had been discovered in Pakistan, and although it was still in the process of being cleaned of rock matrix and reconstructed, it naturally was the focus of considerable interest. "I discussed it with Peter Andrews and Alan Walker, and we went over the oranglike features we could see at the time," recalls Simons. "I was very struck by these resemblances and wondered whether I should jump onto this thing and come out with a strong statement that *Sivapithecus* is like an orangutan. But it wasn't my material. It hadn't been fully described. It was up to David to do that."[41]

Less than a month before the Yale gathering, Simons had been reiterating his conviction of an early origin of the human line before an international symposium in the marble halls of Britain's Royal Society. "It still seems to me that a split-point date in the Miocene around 12–15 million years Before Present is more probable than a mid-Pliocene date of 4–5 million years,"[42] he said after critically reviewing the molecular approaches. "One thing is clear. Strong advocates of the accuracy of 'molecular clocks' have shown a distressing lack of rigour in answering a sequence of problems posed by the series of seemingly late split points calculated generally throughout the family tree of primates. . . . All these are discordant with the dates determined by immunochemical distance."

Time passed and Simons continued to ponder the new developments, but was distracted by his duties at the primate center and by annual field trips to the Fayum Depression in Egypt, where he and his colleagues had been finding spectacular fossils of very early apelike creatures. Meanwhile, however, Richard Kay, a former student of Simons, was preparing a long paper reviewing the hominid status of *Ramapithecus* and *Sivapithecus*, which was eventually published, with Simons as coauthor, in 1983 in a large volume titled *New Interpretations of Ape and Human Ancestry.* Kay had learned well from his supervisor, and was still an enthusiastic supporter of the old party line. The paper's conclusion was straightforward: "Simply stated, ramapithecines are ideally suited to be the ancestors of *Australopithecus* and *Homo.*"[43]

This was Simons' last published support for Prince Rama, and it was more or less a mistake at that. "Of all the papers I've coauthored, this is one I should probably have withdrawn from,"[44] he now acknowledges. "I had very strong misgivings about it." It did lead him to joke to his students and colleagues that "I don't believe Rich Kay is right, but if he is I shall have been vindicated."

Simons finally abandoned his two-decade-long defense of *Ramapithecus* during a December 1982 meeting at Harvard with Pilbeam, who had recently moved there from Yale. By this time both Pilbeam's and Andrews' *Nature* papers had been published, but more important, Steve Ward had now completed his analysis of the GSP 15000 face. He had identified seven or eight features of the palate and face that unmistakably and diagnostically allied it with the orangutan. No question about it. The two men spent a long time alone in the laboratory, talking about the now fully reconstructed face, with Pilbeam explaining precisely what Ward had found. Their discussion recalled times back at Yale, with communication and agreement flowing readily between them. "Eventually I said to David, 'This is a convincing tie between *Sivapithecus* and the orangutan,'" remembers Simons. "We both knew what that meant."

Just as these events were taking place at Harvard, there was occurring back at Yale a new development which provides a final twist to the fossils-versus-molecules story. It involved the new application of an old technique to an old problem.

For several years two biology professors, Charles Sibley and Jon Ahlquist, had been using a method of molecular phylogeny to unravel the evolutionary history of the birds of the world. The method they employed is long established and, under the right circumstances, quite powerful. Known as DNA hybridization, it essentially involves comparing the overall structure—not detailed nucleotide sequence—of the genetic material of two species to see how closely they match. Once again the rationale is based on the idea that the longer two genetically related species have existed independently, the greater will be the difference in their DNA. Because the method involves comparison of all the information-carrying genetic material in an organism, and not just one protein or one gene, there is a strong statistical argument that the accumulation of mutations is indeed clocklike. Having achieved what they considered to be encouraging results with the birds of the world, they decided to turn their attention to the evolution of the apes and humans.

Their results are very interesting indeed. Generally in line with ape/human divergence dates obtained by other molecular methods, nevertheless the DNA hybridization figures are consistently slightly older. For instance, instead of a 5-million-year divergence time for humans and the African apes, Sibley and Ahlquist get a time in the region of 7 to 9 million years. But perhaps the most tantalizing element in these results is that whereas most previous molecular analyses have implied an equal relationship between humans, chimpanzees, and gorillas, Sibley's data suggest quite strongly that humans and chimpanzees are slightly closer together than either is to the gorilla. In other words, humans and chimps may have briefly shared a common ancestor following the separation of gorillas; only then did that common ancestor divide to give us and the chimps.

Pilbeam has come to be very sympathetic to the DNA hybridization technique—not, as some cynics have suggested, because Sibley is from Yale and not Berkeley; and not because the divergence dates are just that bit older than those from other molecular methods; but, he says, because of the obvious statistical power of the method. "I am more prepared to accept the concept that whole-genome DNA changes at a uniform average rate than that any single protein does not fluctuate in rate of change."[45] But accepting that any particular method of molecular phylogeny may be helpful to paleoanthropologists is not Pilbeam's biggest shift in position through this whole affair. It is his belief that molecules may in fact be more reliable than fossils for understanding family trees. "It is now clear that the molecular record can tell us more about hominoid branching patterns than the fossil record does,"[46] he wrote recently.

For a scientist brought up in a tradition in which the lumps and bumps on a fossil are considered the only key to the past, this was a dramatic statement, and it goes to the heart of the battle that was fought over *Ramapithecus.* "Morphology seemed more logical than molecules,"[47] says Adrienne Zihlman. "Morphology is what we 'see' in the sizes and shapes of bones and teeth . . . and it has always been given more weight. Paleontologists have always assumed that chimpanzees and gorillas were more closely related to each other than to humans, because they look so much alike." The key issue is the ability correctly to infer a genetic relationship between two species on the basis of a similarity in appearance, at gross and detailed levels of anatomy. Sometimes this approach works, but sometimes it can be deceptive, partly because similarity of structure does not necessarily imply an identical genetic heritage: a shark (which is a fish) and a porpoise (which is a mammal) look similar because they have become adapted to the same environment, not because they are close genetic cousins.

In the case of *Ramapithecus* there were two potential problems. The first was the shark/porpoise trap that faces all anatomical interpretations of relationship, but on a smaller scale. And the second, and much greater, problem was the power of preconceptions, of seeing in the anatomy what you expect to see.

"Contrary to Simons' and my original view, *Ramapithecus* itself does not have a parabolic dental arcade,"[48] says Pilbeam. "I 'knew' *Ramapithecus*, being a hominid, would have a short face and a rounded jaw—so that's what I saw."[49] Pilbeam and Simons were not uniquely guilty of this error. It occurs often, such is the uncertainty of interpreting fragmentary anatomy in fossils.

For *Ramapithecus*, the shark/porpoise trap took the form of the thick enamel on the cheek teeth, a feature it shares with *Australopithecus*. At one point in the *Ramapithecus* affair, this shared thick enamel was regarded as the main anatomical argument for making a direct ancestral link between *Ramapithecus* and the accepted hominid *Australopithecus*. This supposedly unique link meant—or was interpreted to mean—that *Ramapithecus* must therefore, be a hominid too. In fact, thick cheek-tooth enamel turned out to be something that many of the Miocene apes had and not a specialization uniquely shared by hominids. "It is not always possible to work out what kind of characters you are dealing with," warns Pilbeam.

It was for these two reasons that Pilbeam—and Andrews, for that matter—came to have reservations about the overall power of fossil evidence as compared with data from molecular techniques.

Sarich expresses the dilemma this way: "I know my molecules had ancestors; the paleontologist can only hope his fossils had descendants. In other words, the modern material—be it anatomical, molecular, or behavioral—is immediately relevant to a reconstructive effort; placing the fossils is a much more tricky proposition."[50] He doesn't suggest that molecules are the only source of information, however. "One doesn't want to be foolish and say that the fossil record contributes nothing. But it is true that what it does contribute has been enormously overstated."[51]

But Sarich suspects that there are also other issues in the great *Ramapithecus* debate, which would rate as a modern version of Gregory's pithecophobia. "As I see it, the basic problem has nothing to do with the evidence, be it molecular or paleontological, but with the difficulty most of us have with accepting the reality of our own evolution," he suggests. "We have developed sufficient intellectual maturity to make overt denial of the fact of human evolution impossible. Its positive acceptance, however, is made easier in direct proportion to the distance in time which separates us from our proposed ancestors. . . . This attitude is reinforced by the lure of the 'earliest human.' " This pithecophobia, if it truly exists subliminally within us, can only be exacerbated by the possibility that chimpanzees are closer to us than they are to gorillas.

Simons' interpretation of events is, however, quite different. Preconceptions, he insists, played no part in the misdiagnosis of *Ramapithecus*. "This is supposed to be one of those nice stories in science, that Simons had all these prior conceptions that made him reconstruct the jaw the way he did. And then some clever chaps come along and show that the reconstruction is all wrong and, from other kinds of evidence, that *Ramapithecus* isn't a hominid anyway,"[52] he says. "*Ramapithecus* is disqualified as a hominid because it is like *Sivapithecus*, and *Sivapithecus* has oranglike features. Not because of the biochemistry. And not because the reasons that it looked like *Australopithecus* are not there."

Simons argues that people have been misled by the terms used to describe the shape of the jaw and have not properly examined the reconstruction itself: "If you look at the reconstruction in the 1961 paper it might appear to be a semicircle, but if you draw lines through the axes of the tooth rows you make a V-shape. And the angle of the V is the same as you get from the lower jaws." Pilbeam disagrees, and says that the front of the jaw was made to look like an arc, not like the blunted end of a V-shape. "David feels he got his fingers burned in all this," counters Simons. "I don't think I got my fingers burned. I don't think I saw things that are hominidlike in *Ramapithecus* that aren't there."

In reviewing the *Ramapithecus* affair recently, Milford Wolpoff concluded that it demonstrated how well the science works. "In all, the historical development of human-origins theories and ramapithecine interpretations presents a satisfying contrast to the Piltdown fiasco and reflects the scientific aspect of paleoanthropological studies in a most positive manner."[53] To which Adrienne Zihlman and her University of California colleague Jerold Lowenstein replied: "Unlike Wolpoff, we are struck by the parallels rather than the contrasts between the Piltdown and *Ramapithecus* histories. In both cases, a large number of paleoanthropologists accepted a new 'human ancestor' on the basis of shaky dental and gnathic [jaw] evidence. In both cases, the controversy between believers was resolved by biochemical evidence." And in both cases, as has happened so often in paleoanthropology, practitioners "saw" what they expected to see.

Simons, not surprisingly, agrees with Wolpoff. "I think it is a triumph of anatomy that we have been able to show with better fossils where these animals lie,"[54] he says. "Washburn may have turned out to have been right. But he is one of those fortunate people whose unfounded obsession is right, but for the wrong reasons.

"If I had any preconceptions, it was that the fossils, fragmentary as they were, would tell a story, but not any particular story. In this particular case, I was proved wrong. That's all."

In this particular case, it is true that the formal revision of a major professional paradigm—replacing the idea that *Ramapithecus* was the first hominid with the acceptance that, no, it was not a hominid at all—can be presented as the outcome of objective analysis of new fossil evidence. The *Sivapithecus* faces from Pakistan and Turkey certainly make a convincing story. It could hardly be otherwise, however, for the business of the profession involved in making pronouncements on the course of human history is that of interpreting fossils. That is its stock in trade. Scientific papers by paleoanthropologists are therefore likely to talk about fossils, not molecules—or anything else, for that matter. But it is equally true that the shape and timing of the human family tree which the paleoanthropologists now derive from their fossils is essentially that which Sarich and Wilson proposed twenty years ago on the basis of their molecular clock, and for which they were laughed to scorn.

"At least we've been consistent through these last two decades," says Sarich. "We started with a hominid/pongid divergence date of about five million, and it's been about five million ever since. Paleoanthropologists haven't been consistent, simply because you can't be completely certain what the anatomy means in terms of genetic relationships."[55]

Pilbeam agrees—up to a point, at least. "I am less sanguine than I used to be about the extent to which fossils can inform us about the sequence and timing of branching in hominoid evolution."[56] he told a recent scientific gathering. "I have become convinced that fossils by themselves can solve only parts of the puzzle, albeit important parts." It goes without saying that you need fossil evidence to try to infer what animals now extinct looked like and how they might have behaved. And the anatomical evidence frozen in fossils can give indications of how one creature might be related to another, depending on what features they uniquely share. But even when fossil evidence does reliably indicate a genetic tie between two species, such evidence is less secure when it comes to estimating the closeness of the relationship. "Yes, that's true," concedes Pilbeam. "You are much better off using molecular evidence if you want to be sure about the location and timing of branching points. And that is a difficult admission to have to make for someone who was brought up to believe that everything we needed to know about evolution could be got from the fossils."[57]

Pilbeam says that for him, the molecular evidence eventually began to influence the way he viewed fossil evidence, and vice versa. "There was a kind of rolling feedback between the two"[58] is the way he now describes it. "You can make a good case for saying that if the molecular evidence had not existed, then we wouldn't have recognized the *Sivapithecus* face for what it is. No one can be sure what exactly would have happened in the absence of the molecular evidence, but there's a high probability that the *Ramapithecus* history would have been very different from the way it was."

The clearest message of the *Ramapithecus* affair, however, is the power of preconceptions, which in this case led competent scientists to ignore the evidence of other competent scientists because the conclusions drawn from the evidence were at variance with established ideas. All scientists are guided to some degree by a set of assumptions, usually implicit rather than explicit. "I try hard to detect them in my own thinking," says Pilbeam, "to isolate those assumptions that are not articulated because they are so 'obvious,' yet will seem so silly a few years from now. I am also aware of the fact that, at least in my own subject of paleoanthropology, 'theory'—heavily influenced by implicit ideas—almost always dominates 'data.'. . . Ideas that are totally unrelated to actual fossils have dominated theory building, which in turn strongly influences the way fossils are interpreted."[59]

CHAPTER 7

THE LEAKEYS: FATHER

"It's marvelous,"[1] exclaimed Louis Leakey to his son Richard. "But they won't believe you,"[2] he added with his characteristic, mischievous laugh.

The time was the end of September 1972. Richard had flown back earlier than planned to Nairobi from Lake Turkana in northern Kenya, because he wanted to show a newly discovered skull to his father before Louis left on yet another grueling lecture and fund-raising tour of the United States. Louis, who was already in poor health and under medical supervision for very high blood pressure, had just succeeded in persuading his son Colin to leave Uganda to escape the looming threat of the Idi Amin regime. Exhausted, but much relieved by events, Louis was understandably in good spirits. But when he saw what Richard carefully unwrapped on his desk that morning at the Nairobi Museum, his delight multiplied immeasurably. There before him was the skull known as 1470, a large-brained human ancestor which, at that time, was thought to be nearly 3 million years old.

"It represented to him the final proof of the ideas that he had held throughout his career about the great antiquity of quite advanced hominid forms,"[3] explains Richard. Louis Leakey had spent four decades searching for evidence of early *Homo*, and he was convinced that Africa was the place where he would be found.

Richard's new find seemed to vindicate all that Louis had striven for. "Seeing and handling the '1470' skull was an emotional moment. . . . He was delighted that it was a member of my team who had made the find at a Kenyan site." A few days later, Louis Leakey died of a massive heart attack in London.

Louis Leakey, born in Kenya in 1903, had devoted more than forty years to human-prehistory research in East Africa, much of it with his second wife, Mary. His interests spanned the complete spectrum of prehistory, from recent archeological sites, where he found the most delicate obsidian implements and exquisite rock-shelter paintings, to the earliest evidences of human origins. There is no doubt, however, that his consuming passion was the latter: he wanted to find the first human, and he wanted to find him in Africa. Moreover, as Mary Leakey now recalls, "it was one of his creeds that man went back a very, very long way."[4] Indeed, throughout his colorful and controversial career, the name Louis Leakey became virtually synonymous with the idea of an ancient origin of *Homo*, the line that led directly to modern man. "Louis Leakey was captivated by the idea of early true man,"[5] as Don Johanson recently put it.

Although Leakey had highly respectable credentials—a degree in anthropology from Cambridge University, England, and a fellowship in one of its most revered colleges—he was more explorer than scientist. He loved to be out in the field, discovering new sites and returning to established ones: particularly, of course, Olduvai Gorge in Tanzania. And he was irked by the conservative ways of the scientific establishment. He never held an academic post, and indeed became scornful of bookish armchair scholars. Leakey's paleoanthropological work was something to be fitted in among myriad duties centered around the National Museum in Nairobi and other government activities. Perhaps this distance from academia freed him of the normal establishment constraints, for his claims were often a source of consternation among his colleagues in universities. When he said to Richard that day in September, "They won't believe you," he was pointedly echoing his own experience, and vicariously relishing the prospect of a fight.

Leakey first went to Olduvai Gorge in 1931 while on the Third Archaeological Expedition to East Africa, which he organized from Cambridge. The reason for the visit was to try to solve the mystery of "Oldoway Man," a skeleton discovered by a German scientist,

Hans Reck, in 1913. The mystery was that the skeleton looked to be completely modern, yet according to Reck it had been excavated from deposits more than a million years old. Leakey had seen the skeleton twice before the trip to Olduvai, in 1927 and 1929, while it was being studied by Professor Theodore Mollison in Munich, Germany. Leakey had been unimpressed: "Almost certainly this is not contemporary with the fossil deposits of the gorge in which it was found,"[6] he concluded in 1929. "[It] probably represents an intrusive burial." In other words, Leakey considered that Oldoway Man was a modern human who had been buried in a grave dug into million-year-old sediments, thus giving the appearance that he had died a very long time ago.

Reck, however, remained adamant, and joined the 1931 expedition determined to demonstrate to the skeptical Leakey that the geology would prove his case. During his previous visit to the gorge Reck had spent three months searching unsuccessfully for stone tools, which he felt must be there if his Oldoway Man was genuine. Leakey, who had been collecting stone tools in various sites in Kenya since his teens, offered a £10 wager that he would find tools within a day of their arrival at Olduvai. Reck accepted, and was more than pleased to pay up when within only a couple of hours, Leakey had picked up several splendid hand axes made from basalt lava. The German had overlooked such implements, it transpired, because he had been searching for flint tools of the sort he had known from European sites.

Although it is not documented, it seems a good bet that the discovery of the hand axes had a profound effect on Leakey's outlook. In any case, within a few days of pitching camp at Olduvai, Leakey, Reck, and A. T. Hopwood, another member of the expedition, had dispatched a paper to the British journal *Nature* confirming Reck's original conclusion. Leakey also sent a short article to *The Times* of London claiming that the expedition had established "almost beyond question that the skeleton found by Professor Reck in 1913 is the oldest known authentic skeleton of *Homo sapiens*."[7] Leakey's biographer, Sonia Cole, remarks that Reck "must be one of the few people who succeeded in swaying Louis once his mind was made up."

Here, on the very first visit to the site that was to play such a central role in Leakey's professional life, a pattern was set. He wanted to believe in ancient *Homo*, and so suspended the degree of critical judgment he might otherwise have applied to the evidence.

As it turned out, Oldoway Man was very quickly toppled from his pedestal when others irrefutably demonstrated that the skeleton was indeed an intrusive burial. Leakey agreed, and in October 1934 wrote the following in the preface to his *Stone Age Races of Kenya*: "On the evidence of the Oldoway skeleton there seemed at first to be an indication that the species of *Homo sapiens* was of great antiquity in East Africa, but the investigation dethroned Oldoway man from his proud position of being probably the oldest *Homo sapiens*. Few people were surprised at this, because the Oldoway skeleton not only was truly *Homo sapiens* but a very highly evolved specimen of *Homo sapiens*."

Bloodied but unbowed, Leakey continued: "It is a strange coincidence that the very primitive and generalised men of *Homo sapiens* type—as represented by the Kanjera skulls—should have come from the very same geological horizon from which it was originally held that the Oldoway man was derived."[8] In other words, Leakey was not very much upset about the collapse of Oldoway Man's claim, because he had meanwhile found other evidence, in the form of fragments of several skulls from a site called Kanjera, that were also *Homo sapiens* and were close to a million years old. His claim for ancient *Homo* therefore remained intact.

The Kanjera skulls had been uncovered in March 1932 at a site in western Kenya. At the same time, part of a jaw known as the Kanam mandible was also found at another site very close by. Leakey considered the jaw, like the skulls, to be very close to *Homo sapiens* and also to be extremely old. Judging the mandible to represent a type of human just a little distant from *Homo sapiens*, however, Leakey decided to give it a new name. "While I have formed the opinion that the . . . creation of a new species— *Homo kanamensis*—is justified, I would like to point out that *Homo kanamensis* must be regarded as standing much closer to *Homo sapiens* than do any known other genera or species, and that in all probability *Homo kanamensis* is the direct ancestor of *Homo sapiens*." Not content with this, Leakey added that he believed that the jaw "is not only the oldest known human fragment from Africa, but the most ancient fragment of true *Homo* yet discovered anywhere in the world."

Elwyn Simons describes this as a fine example of the Louis Leakey syndrome, which is a subsidiary, but highly prominent, component of Leakey's pursuit of ancient *Homo*. Leakey's modus, says Simons, was that "The fossils I find are the important ones and are on the direct line to man, preferably bearing names I have coined, whereas the fossils you find are of lesser importance and are all on side branches of the tree."[9] Exaggerated, perhaps, but not entirely unfair.

While the episode over the Oldoway skeleton had not exactly enhanced the young Leakey's reputation at the beginning of his career, developments surrounding the Kanjera and Kanam fossils would prove to be a major embarrassment from which his scientific standing would take a very long time to recover. A lesser man than Leakey would have been crushed by the experience.

Although Leakey at first received encouragement and plaudits for his work on Kanam and Kanjera, particularly from a special meeting of the Royal Anthropological Institute held in Cambridge in March 1933, a subsequent field investigation proved to be his downfall. At Leakey's invitation, Professor Percy Boswell, a geologist from Imperial College, London, went to Kenya in January 1935 to inspect the sites. A series of unfortunate incidents put Boswell "in a bad humour," as Leakey recorded in his field diary for 18 January. Not only was Leakey unable to locate the exact spot from which the Kanjera skulls were recovered, but the iron pins that had been placed to mark the site of the Kanam mandible had apparently been removed by a local fisherman who wanted the metal for fishhooks. Worse, the photograph of the Kanam site that Leakey had used in an exhibition at the Royal College of Surgeons, London, in January and was due to publish in his *Stone Age Races of Kenya* turned out to be of a location several hundred yards distant. On discovering this last fact during the field trip, Leakey was compelled to cable Oxford University Press to hold up delivery of the book so that an erratum slip could be inserted.

Boswell, a scientific pedant and a stickler for detail, was distinctly unimpressed, and subsequently wrote a damning paper which was published in *Nature* on 9 March 1935. An unfavorable report was also delivered to the Royal Society. Boswell may well have had his own, less direct reasons for criticizing Leakey so stridently. An ardent supporter of Piltdown, he would have found it most uncongenial to have to accept fossil men of even more modern appearance but of the same geological age as Piltdown—and in Africa, not England.

Furious with Boswell, Leakey drafted a lengthy reply, which *Nature* rejected. A shorter version finally appeared early in 1936. Nevertheless, the damage was done: Leakey was tarred with a reputation for being less than meticulous in his scientific practice. Dr. A. C. Haddan, a friend and colleague from Cambridge, wrote to Leakey on 21 March 1935, "I must confess that I am disappointed at the casual way in which you dealt with the matter . . . It seems to me that your future career depends largely upon the manner in which you face the criticisms."

Leakey faced the criticisms as he always did: pugnaciously. But he was wrong, on both counts. The Kanjera skulls were eventually shown to be a mere 15,000 years old. And the Kanam jaw, while old, had been distorted by a pathological growth that had given it a *Homo sapiens*–like appearance.

This latter issue—the great antiquity of man—so dominated Leakey's view of the past that it would lead him repeatedly to see in fossils what he wanted to see. The reasons for this are several and have to do with his family background, his intellectual associations, and perhaps even his deeply held religious beliefs.

Leakey had studied anthropology at Cambridge in the late 1920s, when an important change had been going on in the profession. In 1871, Charles Darwin had predicted that Africa would prove to be the cradle of mankind, primarily because man's closest relatives, the chimpanzee and gorilla, live there now. The idea remained current until the end of the century, when Eugène Dubois discovered a primitive form of man in Java, which he called *Pithecanthropus* (now *Homo*) *erectus*. Gradually from then on paleoanthropologists began to think more of Asia as man's birthplace, but the change was by no means immediate, as many authorities rejected Dubois's claims. Henry Fairfield Osborn's enthusiasm for Central Asia rode on the rising swell of opinion, as we saw in an earlier chapter. But what really swiveled people's eyes from Africa to Asia was the discoveries from 1926 onward of Peking Man, another form of *Homo erectus*.

So by the time Leakey had finished his anthropological studies at Cambridge and was eager to get into the field, Asia, not Africa, was considered the place to look for the earliest human forms. When he told a Cambridge professor that he planned to go to East Africa to look for human fossils, he was told, "Don't waste your time. There's nothing of significance to be found there. If you really want to spend your life studying early man, do it in Asia." To which Leakey replied: "No. I was born in East Africa, and I've already found traces of early man there. Furthermore, I'm convinced that Africa, not Asia, is the cradle of mankind."[10] Leakey's early experience had, it's a fair guess, once again been important. A keen observer and naturalist, he had as a young boy turned his attention from ornithology to the search for stone artifacts and other archeological materials. In an interview toward the end of his life, he said: "[I] decided at the age of thirteen to find out if Darwin was right. I was born in Africa and I became excited with the idea that possibly everyone was looking in the wrong place."[11] Leakey's idea of intellectual fun always was to prove everyone else wrong.

The "where" of Leakey's search for human origins is therefore probably readily explained: he was born in Africa; he had already found stone tools there; and the establishment thought Asia was the place to look. But what of the "when"? Why did he consistently seek—and find—traces of early man in ancient geological formations? Was it, as Michigan paleoanthropologist C. Loring Brace has put it, related to "the tradition established in British anthropological circles of being so anxious to prove the great antiquity of *sapiens* forms that any such indication, no matter how tenuous, will be accepted until proven false"?[12]

Leakey was indeed educated in the British anthropological tradition, at a time when the influence of Piltdown Man on the profession's intellectual outlook was at its height. The principal effect of Piltdown had been to encourage people to believe that modern human forms were already established early in the geological record. And the chief proponent of Piltdown in England was Sir Arthur Keith, Leakey's mentor for several years. Leakey frequently took fossil material to Keith's laboratories at the Royal College of Surgeons in Lincoln's Inn Fields, London, and the two men spent many hours discussing specific and general issues in their science. They even used the same drawing apparatus for some of their publications. So, was the British school, and specifically Keith, the source of Leakey's devotion to the ancient-origin idea? A comparison of the two men's writing is instructive.

For Keith and his professional contemporaries the antiquity of man was a major preoccupation, and in 1915 Keith published a magnificent two-volume work on the subject titled simply *The*

Antiquity of Man. In it he wrote that: "When we speak of the antiquity of man . . . most of us have in mind not the date at which the human lineage separated from that of the great anthropoids, but the period at which the brain of man had reached a human level."[13] This fascination with man's big brain was, of course, what laid the ground for the eager and uncritical acceptance of the Piltdown hoax. Leakey was entrapped by this fascination too, explicitly so in his early professional years and then later more as an implicit underlying theme.

Keith's rationale for arguing for an early origin of man, in the above sense, was simple and direct. "I grew up under the belief that evolution proceeded in a leisurely manner and required long stretches of time to work her effects,"[14] he wrote in 1925, "a belief I still cling to." The first part, then, was that evolution was a slow process. Accepting this, and believing, as Keith and his contemporaries did, that the modern races of man represented considerably differentiated types, one finds the second part was obvious: "I do not think that any period less than the whole . . . of the Pleistocene period . . . is more than sufficient to cover the time required for the differentiation and distribution of modern races of mankind."[15] In other words, big-brained man must have an ancient origin because it must have taken a very long time to progress from this ancestral stock to the widely separate races that exist today. "The wolf, the bear and . . . the gibbon . . . had reached their present stage of evolution in the Pliocene,"[16] Keith noted, "and if this were possible for them, why deny the same possibility to *Homo sapiens*?"

Leakey's line of argument, expressed first of all in his *Adam's Ancestors*, published first in 1934, was identical. "I would say . . . that we have learnt that evolution has been very much slower than we have sometimes been led to believe,"[17] he notes. "The subdivision of a species into a number of distinct races is a slow and gradual evolutionary process, for which ample time must be allowed."[18] Once again, the slow rate of evolution means that modern racial differences were a long time in the making. Leakey goes on to argue that "The presence of four completely different types of man at the beginning of the Pleistocene [he is referring to Java Man, Peking Man, Piltdown Man, and Kanam Man] suggests to me that the common ancestor must be looked for in deposits at least as old at the Miocene period."[19] This, again, is an echo of Keith, who wrote a few years earlier that "There is not a single fact known to me which makes the existence of the human form in the Miocene period an impossibility."[20]

In constructing his versions of the human evolutionary tree, Leakey always produced a very bush-like structure, with very few species actually falling on the direct line to modern man; most were twigs that led to dead ends. "Peking man, Java man, and the Neanderthaloids . . . are really nothing but various aberrant and overspecialized branches that broke away *at different times* from the main stock leading to *Homo*,"[21] he said. None of these "was, on the evidence available, in any way ancestral to *Homo sapiens* (in spite of some opinions to the contrary), for all show specializations which were once thought of as 'primitive' characters, and which led to the belief that these types represented 'primitive' stages of man, not highly specialized offshoots of the human stock."[22] This habit of pushing more or less every fossil type onto a side branch, damned through being too "specialized," was again a clear echo of Keith's approach. The trees they drew looked very similar to each other, both in style of draftsmanship and in form.

Leakey, incidentally, accepted the authenticity of Piltdown in the 1930s, and, like his mentor, hung it on yet another dead-end twig. "As the evidence is at present," he wrote in 1934, "the Piltdown man must be considered as having been more or less contemporary with the Kanam man, and therefore could not be regarded as ancestral to it. Nevertheless, the Piltdown skull is probably more nearly related to *Homo sapiens* than to any other yet known type, and it must be regarded as a rather primitive side-branch of the same stock."[23]

There would therefore appear to be no mystery in why Leakey began his search for human origins with the orientations he so clearly had. He was continuing Keith's tradition. But what is interesting is the fact that he stayed fast with this tradition—the essence of it, at least—while the academic colleagues around him were modifying theirs. This is where the mystery lies.

There is an interesting footnote to Leakey's devotion to Keith's tradition which is at the very least a striking coincidence. As it happens, Leakey's enthusiasm, albeit brief, for the antiquity of *Homo sapiens* as evidenced by Oldoway Man was a precise parallel with yet another aspect of Keith's career. Long before Piltdown came along, Keith was convinced of the early origin of large-brained ancestors because of Galley Hill Man, a modern-looking skeleton dug from the early Pleistocene gravels of the River Thames, near London, in 1888. In his many investigations of the skeleton, Keith identified numerous primitive anatomical details, which, he argued, proved its antiquity. It eventually transpired, however, that Galley Hill was a modern skeleton, which, like Oldoway Man, had the geological association of being older because of intrusive burial. Keith therefore finally had to abandon his favorite skeleton; but of course he had Piltdown to fall back on for support of his theories, just as Leakey had Kanjera and Kanam.

Leakey labored tenaciously in his search for human ancestors, and by the end of the 1950s he believed he had fossil evidence for the modern great apes and gibbons going back some 20 to 30 million years. He therefore estimated that the apes and humans "may have come from a common source, probably more than 40 million years ago."[24] Like Elwyn Simons and David Pilbeam, as reported in an earlier chapter, Leakey was drawing long straight lines through time, linking ancient fossils as direct ancestors to living species. It was a common theme of the time. As his research progressed, Leakey became less entranced with the search for early *sapiens* and more focused on the beginning of the human form itself, in which, nevertheless, the large brain was emphasized. This emphasis led to the ultimate irony in 1964 when, in order that a new fossil of his should be admitted as a species of *Homo*, he modified the definition of the genus to include smaller-brained animals.

The discovery in 1961 of the 14-million-year-old *Kenyapithecus wickeri*, and a few years later of even more ancient *Kenyapithecus africanus*, was important in Leakey's saga, because it provided for the first time what he believed was an identifiable hominid: "the oldest member of the Family Hominidae known at the present time,"[25] he noted in January 1967. But undoubtedly the major turning point in his professional life resulted from the discovery in 1959 by Mary Leakey of the famous fossil skull named *Zinjanthropus*, meaning East African Man. Here, at last, announced Leakey, was an ancient, primitive form of man that was directly ancestral to modern humans. But the admission of "Zinj" onto that sacred direct path to man posed a few problems for Leakey, which he nevertheless surmounted: they were problems of how one recognizes really primitive man, and how fast evolution can transform such a primitive form into modern man. Zinj, incidentally, was what made Leakey world-famous, and with that fame came a more secure source of research funding than he and Mary had ever previously enjoyed.

By the time Mary uncovered Zinj at Olduvai Gorge on that now-famous July day of 1959, researchers in South Africa had culled a large haul of australopithecines from the Transvaal caves, some of which were like Dart's original Taung child while others were considerably more robust. The Taung-like fossils were known as *Australopithecus africanus*, while the bigger ones had been named, appropriately enough, *Australopithecus robustus*. Leakey visited Johannesburg and Pretoria in 1945 to view the fossils, and immediately formed strong opinions. "He talked about the australopithecines a lot on his return," remembers Mary. "But he didn't believe for a minute that they were ancestral to *Homo*. That was anathema to him. He thought it was anatomically impossible, that they were just too specialized."[26] Dart and his Taung child, incidentally, did not rate a mention in Leakey's 1934 edition of *Adam's Ancestors*, in which respect Leakey was in line with most contemporary authors.

Many people called the australopithecines "man-apes" or "ape-men," to which Leakey vigorously objected. "It implies that these creatures . . . represent a missing link between apes and man. There are some scientists who believe this to be the true explanation, and if they are right then 'man-apes' or 'ape-men' may be the correct term to use, but I prefer to call them 'near-men,' for it seems to express their status much more accurately,"[27] he wrote in 1953. "They represent a very aberrant and specialized offshoot from the stock which gave rise to man. In many respects, however, they certainly stand much nearer to man than they do to any of the living great apes, so that the term 'near-men' seems to fit them well."[28]

When Leakey rushed to see Mary's discovery of Zinj on that famous July day of 1959, his first comment, she remembers, was "Oh, dear. I think it's an australopithecine."[29] His first assessment was, as it later transpired, correct. Zinj was an even larger version of the *Australopithecus robustus* species known from South Africa. It had a massive bony crest running down the middle of its head, to which the muscles of the lower jaw were anchored; its face was curiously dished; the cheekbones flared like great flying buttresses; the cheek teeth were enormous and the front teeth tiny. A most striking and bizarre physiognomy, to be sure. Nevertheless, Leakey quickly persuaded himself that Zinj was significantly different from the South African 'near-men' and showed enough similarities to *Homo sapiens* to merit designation as a direct ancestor to modern man. Within a month, he had a paper published in *Nature* proclaiming East African Man's manhood.

A decisive factor in his judgment was the association with stone tools, as it had in a sense been with the long-forgotten Oldoway Man. Mary and Louis had for decades been making collections of stone artifacts from Olduvai, and Mary had revolutionized archeology with her classification of these early stages of tool technology. As it happened, Zinj's skull was lying on what appeared to be some kind of living floor amid a scatter of the most primitive stone tools and many broken-up animal bones. Given similar circumstances at sites elsewhere in the world, Leakey had often suggested that the fossil hominid on the site had clearly been the victim of some more advanced man, whose remains had not yet been found.

The interpretation with Zinj, however, was different, he argued. "There is no reason whatever, in this case, to believe that the skull represents the victim of a cannibalistic feast by some hypothetical more advanced type of man."[30] he wrote in *Nature*. His line of logic was that "since [the skull] was not smashed up [before fossilization], while the bones of all the other animals are broken, it may reasonably be assumed that the skull represents the maker of the culture who lived upon the floor. In view of our accepted definition, therefore, of man as a 'primate who makes tools to a set and regular pattern,' or 'Man—the Tool Maker,' we must accept *Zinjanthropus* as a true 'man.' "[31]

Persuaded by the archeological evidence that Zinj is indeed 'true man,' Leakey then asks rhetorically, "Does he fit in with our concept of what man should look like?"[32] His answer is "Hardly," and for good reason. "At first sight, this new skull very strongly recalls the 'ape-men' . . . that Broom and Robinson found at Swartkranz, in the Transvaal. . . . But when the skull is closely compared with the Swartkranz specimen known as *[Australopithecus robustus]*, it is found that the differences are far more numerous—more significant than the resemblances." The differences Leakey identified include the curvature of the cheek region, the structures around the ears, and the base of the skull. These, he said, allied *Zinjan-*

thropus with *Homo sapiens.* He even ventured to predict that "I should indeed be surprised if the lower jaw, when we find it, does not exhibit the form characteristic of speaking man."[33]

When Leakey announced Zinj to the world in the pages of *Nature*, he caused some wry amusement among his colleagues by avowing that "I am not in favour of creating too many generic names among the Hominidae [the human family], but I believe that it is desirable to place the new find in a separate and distinct genus. I therefore propose to name the new skull *Zinjanthropus boisei."* Leakey, of all people, was a supersplitter.

Formally, he had to base his diagnosis on the anatomy of the skull, not on its cultural association, even though this latter was clearly the key factor in his mind. According to his longtime friend and colleague F. Clark Howell, Leakey had arrived at the list of anatomical similarities with *Homo* and differences from the robust australopithecines even before he had performed a direct comparison with the original specimens.

"I was the first scientist to see Zinj, apart from Louis and Mary," recalls Howell. "I had dinner at their house shortly after they brought Zinj back from Olduvai. Louis hadn't said anything about the discovery at first, but after dinner he brought out a big cookie box, placed it on the table, opened it, and said, 'There—what do you think of that?' Louis liked to do those kinds of theatrical things. Mary just sat there with a nice smile on her face. I told Louis I thought it was just like a robust australopithecine, and he said, 'No, no, no,' and proceeded to explain to me in great detail why he thought I was wrong. Then he said he was going to go to South Africa to compare Zinj with the original specimens of the robusts. Louis already had a draft of a paper on Zinj completed at that point."[34]

In retrospect, it is easy to see that Leakey's eagerness to find an early ancestor to 'true man' had led him to overinterpret the anatomical evidence. It also got him tied up in a definition of man that included stone-tool culture; this latter would come back to haunt him. But in accepting Zinj as a true human ancestor, Leakey made for himself another problem, which once again he circumvented. Specifically, this had to do with the rate of evolution among hominids. Leakey, remember, considered evolution to be a slow process. He therefore found himself faced with a dilemma, as he explained in a lecture in Cape Town to the South African Archaeological Society in 1960: "If we are willing to concede the

possibility, as I most certainly am, that the genus *Homo* was derived from an australopithecine resembling *Zinjanthropus,* then
we have to ask ourselves whether or not a sufficiently long time
interval elapsed between the Lower Pleistocene and the closing
stages of the Middle Pleistocene for the transformation, by evolution, from *Zinjanthropus* to something resembling *Homo.*"

When *Zinjanthropus* was first unearthed, it was considered by
current geological estimates, to be some 600,000 years old, which
would have been a very short time for the evolutionary change
Leakey envisaged. Almost immediately, however, the application
of radiometric dating to sediments at Olduvai Gorge revealed the
new fossil to be more than three times the original estimate: to be
specific, 1.75 million years. This unexpected discovery certainly
helped out the time dilemma, but Leakey had an even better explanation. He reminded his audience at the South African Archaeological Society meeting that animals under domestication
demonstrably evolve at must faster rates than under natural conditions. He then went on to say, "Do we not, too often, overlook
the fact that as soon as man began to make tools to a set and
regular pattern, as *Zinjanthropus* was in the Lower Pleistocene,
man had, in fact, set up conditions of domestication *for, himself.*
From that moment onwards he potentially accelerated the results
of the natural evolutionary process within his own stock." In other
words, *Zinjanthropus,* by his adoption of a stone-tool culture, propelled his own evolution toward *Homo sapiens* at an ever-faster
rate. "Once he had become the maker of stone tools, there is no
reason at all why human evolution should not have been as rapid
as that of his many domestic animals."

Leakey's enthusiasm for Zinj's status as a direct forerunner of
man encouraged him to suggest that perhaps the Kanam mandible
wasn't *Homo kanamensis* at all, but instead was a female *Zinjanthropus.* Most of the paleoanthropological establishment were,
however, more than a little skeptical of Leakey's various claims,
including the idea that Zinj was anything other than an australopithecine, and that he was on the direct line to man.

As events transpired, Zinj was to be rapidly deposed; but on this
occasion it was by Leakey's own hand that it came to pass. A
statement he made in September 1960 turned out to be hauntingly
prophetic: "It is precisely by his manufacture of the first known
pattern of implements that I believe *Zinjanthropus* can claim
the title of earliest man—at least until other, more distant, tool-

makers are found."[35] Two months later, the first significant fragments of that other tool-maker turned up at Olduvai Gorge, right at the spot where Zinj had been found. The eclipse of East African Man's claim to be the first true man was to be embarrassingly rapid and complete.

In November 1960, Jonathan Leakey, Louis' eldest son with Mary, was prospecting very close to the Zinj site and found parts of a cranium and the lower jaw of a hominid that had died at about twelve years of age. To judge from the geology, it appeared that the hominid had lived even earlier in time than Zinj, and was therefore called the pre-Zinj child. But more important, the anatomy appeared to suggest that the child was a different type of hominid from Zinj: superficially, it was of a much lighter build, and the skull fragments seemed to portend an expanded brain. Over the next three years more and more bits and pieces of this new type of hominid were recovered, including parts of a foot, a hand, and ribs. At one point Leakey wrote to a colleague in England saying that he and Mary appeared to be moving toward a whole skeleton; but it was not to be.

Leakey was quickly convinced that here was something different from *Zinjanthropus*—something entirely new to paleoanthropology, and something very close to the earliest *Homo*, his ultimate goal. And eventually he would, in a press release, urge his colleagues "to review all their previous ideas about human origins and to substitute for those theories new ones which would be more in keeping with the facts that were now known."[36] But for more than three years he simply published formal descriptions of the various fossil finds and said nothing publicly about his interpretation of them. This time he was going to be much more cautious. This time he wanted to assemble as strong a case as he could for what would be a major event in paleoanthropology. Nevertheless, when it finally came, on 4 April 1964, Leakey's announcement provoked a storm of protest, reverberations of which continue to this day.

Throughout the period of the excavations and steadily rising excitement, Leakey maintained a regular correspondence with Sir Wilfred Le Gros Clark, in Oxford, England, with whom he had been a friend and colleague for many years. Le Gros Clark was Britain's leading paleoanthropologist, and his reaction to Leakey's eventual pronouncements on the pre-Zinj child was to be crucial in setting the tone of the ensuing debate. Less than two weeks after Jonathan's discovery of the skull fragments, Leakey wrote of it to Le Gros Clark; "The morphology simply does not fit into the australopithecine pattern,"[37] he suggested. "I have an increasingly strong feeling that . . . we have two entirely distinct hominids living side by side in Bed I at Olduvai, just as we have eight different genera of pigs side by side, and so on." On the same day Leakey also wrote to Phillip Tobias, who was working on the Zinj skull in Johannesburg, and to Dr. M. W. Sterling, at the Smithsonian Institution in Washington, D.C., telling them that he was practically certain he and Mary now had a new hominid. Both were asked to keep the information confidential.

Less than a month later, on 7 December 1960, Leakey wrote again to Le Gros Clark, this time a six-page epistle giving detailed descriptions of all the fossils and enclosing photographs. "To me there is no doubt at all, the more I study the specimen, that we are dealing with a true hominid with certain strongly marked very primitive characters, which was slightly earlier on this site than *Zinjanthropus*, but presumably also contemporary with him," said Leakey. Then, identifying a topic that later was to be at the eye of the storm, he added: "This will raise a major issue as to whether the Oldowan culture was made by the Australopithecine *Zinjanthropus*, or by this quite distinct and more manlike, though very primitive, hominid."

To Leakey's evident frustration, Le Gros Clark remained unimpressed with his arguments for the new fossils, and throughout 1961 continually urged a more conservative approach. In June, Le Gros Clark wrote, saying that from the evidence he and his colleagues had seen, he considered the fossils to be inseparable from *Australopithecus*. And on 5 July he responded to a rather piqued Leakey, "I do hope that you will not commit yourself too strongly at this stage." But it was already too late, for Leakey was by this time firmly set in his view.

Meanwhile, Leakey had recruited Phillip Tobias' help in describing the pre-Zinj fossils, in addition to working on the East African Man's skull itself. Tobias now held the professorial chair once occupied by Raymond Dart, and had worked with Dart for a very long time. Professionally, Tobias was therefore closely associated with *Australopithecus africanus*, and, initially at least, was reluctant to accept Leakey's suggestion that the pre-Zinj child was something different. On 1 May 1962, for instance, Tobias wrote to Leakey that "My present feeling about the child is that it is an

australopithecine." As an attempt to somehow soften his statement he offered that he was merely "thinking aloud." In reply, on
7 May, Leakey said, "If you finally decide that the child, on all its
characters, is an australopithecine, you must clearly say so, but I
shall then reserve the right to say the opposite if I find your conclusions unconvincing." He also displayed his growing impatience
with sitting on what he considered irrefutable evidence, by adding
that "Mary and I are seriously considering whether it would not
be wise to name the pre-Zinj child now."

Tobias equivocated, and Leakey applied ever more pressure on
his colleague. "Mary and I are sure (more and more so every time
we go over the data) that it is NOT *Australopithecus*," he wrote
on 28 December 1962. "I think only those with 'Psychosclerosis,'
as Le Gros once put it, and who cannot bear the idea of two contemporary branches of hominid, could ever put it in that sub-
family." Leakey knew that his colleagues would react negatively
to his claim that two different hominids were living cheek by jowl,
because such an association was counter to prevailing paleoanthropological theory and no one had previously challenged it directly with fossil evidence. "The part played by prejudice in
relation to scientific controversy was very strong in the thirties
and remains so to this day,"[38] Leakey noted a little later.

Tobias continued his resistance late into 1963, but was finally
convinced, he says, "by a whole spate of additional specimens that
Mary found in October that year."[39] From the five or six individuals now represented in the sample, there were sufficient anatomical features from which to conclude that the hominid was indeed
different from Zinj, and, most crucial of all, that it was *Homo.* "It
was at this point that I changed my mind," says Tobias. The persuasive aspect of the evidence was that "It became plain that this
Olduvai group of fossil hominids had a mean [cranial] capacity that
was nearly 50 per cent greater than the mean of *A. africanus*,"[40]
he noted recently.

"Phillip took a lot of persuading," remembers Mary Leakey.
"Louis had to bludgeon Phillip to convince him. No one lightly
names a new hominid species. But Louis loved it."[41]

Shortly after Tobias was won over, Leakey wrote again to Le
Gros Clark, on 6 January 1964, telling him that "I shall be publishing a scientific name and diagnosis in association with Phillip
Tobias in 'Nature' in the near future but I thought you would like
to know beforehand." Unable to contain his excitement at the

prospect of its impact, he said that "The net result, of course, means that we will have to look for a common ancestor of hominines and australopithecines way back in the Pliocene." In short, because primitive 'true men' and 'near-men' were clearly coexisting at Olduvai about 2 million years ago, the common ancestor of the two lines must have originated much further back, perhaps some 10 million years ago (according to the geological time scale then current). This would make true *Homo* a very ancient stock indeed.

There are very strict rules of convention that must be followed when a scientist names a new species of an existing genus. These involve a careful description of the new specimen to show how the animal fits in with the definition of the genus (in this case *Homo*) and how it differs from other closely related species. The naming of a new species of hominid has always been likely to touch off a good deal of lively discussion within the profession, so Leakey was certainly guaranteed that much. But the fact that at the same time he had to adjust the definition of the genus *Homo* in order to accommodate his new species transformed what might have been properly scholarly jousting into near outrage. "We were berated by a couple of our colleagues for modifying the definition of *Homo*, as though it were as sacrosanct as the law of Medes and Persians,"[42] remembers Tobias.

"We have come to the conclusion that, apart from *Australopithecus (Zinjanthropus)*, the specimens we are dealing with from Bed I and the lower part of Bed II at Olduvai represent a single species of the genus *Homo* and not an australopithecine," concluded Leakey and Tobias in the now famous 4 April 1964 issue of *Nature*. The two men had also been joined by a third author, John Napier of London University, whose expertise on hand and foot anatomy had been sought to aid in the diagnosis. "But if we are to include the new material in the genus *Homo* (rather than set up a distinct genus for it, which we believe to be unwise), it becomes necessary to revise the diagnosis of this genus."[43]

Previous diagnoses of *Homo* had focused on the so-called cerebral rubicon: a hominid had to exceed a certain brain size in order to be judged worthy of full manhood. The trouble was that different authorities put the threshold at different levels. Sir Arthur Keith, for example, put it at 750 cubic centimeters, whereas for Henri-V. Vallois, a larger figure of 800 c.c. was necessary, while Franz Weidenreich was content with 700 c.c. Clearly, it was all rather arbitrary. Nevertheless, Tobias had measured the brain size of the pre-Zinj child to be 675–680 c.c., which did not qualify on anyone's scale. In fact, it fell midway between those of the largest australopithecine and the smallest *Homo erectus*, which until then had been a no-man's land for hominids.

The diagnosis offered by Leakey, Tobias, and Napier was more comprehensive, and included habitual bipedal posture and gait, a precision grip, and a brain capacity much smaller than previously proposed. The new cerebral Rubicon was to be 600 c.c., which figure just admits the pre-Zinj child into the *Homo* fold. Although the association with stone tools did not formally enter into the diagnosis, there was a clear implication that it had been important in the authors' conclusions. "When the skull of *Australopithecus (Zinjanthropus) boisei*," they noted, "was found on a living floor at FLK I, no remains of any other type of hominid were known from the early part of the Olduvai sequence. It seemed reasonable, therefore, to assume that this skull represented the makers of the Oldowan culture. The subsequent discovery of remains of *Homo habilis* in association with the Oldowan culture at three other sites altered the position." They accepted the possibility that both Zinj and the pre-Zinj form were stone-tool makers, but concluded that "it is probable that the latter was the more advanced tool maker and that the *Zinjanthropus* skull represents an intruder (or victim) on a *Homo habilis* living site."

Soon afterward, Tobias and Napier, responding to criticism from Bernard Campbell, a young anthropologist from Cambridge, England, wrote a letter to *The Times* in which they leaned even more heavily on the cultural association as support for the *Homo* status of the pre-Zinj child. "On the basis of our comparisons, we conclude that the anatomy of the new fossils is about midway between that of *Australopithecus* and of *Homo erectus*. But anatomy alone could not tell us whether the new creature was the most advanced *Australopithecus* or the lowliest *Homo*. The answer was provided by a strong body of evidence that the *habilis* man was the maker of the early stone tools." This statement caused Leakey some embarrassment, and he felt compelled to dissociate himself from it publicly. Tobias now defends his and Napier's actions by arguing that "Our importing of cultural evidence more strongly into the scale-pans in weighing up the generic status of *H. habilis* was entirely in keeping with accepted procedure that ethnological evidence may be added to morphological evidence in the assessment of the systematic status of a group."[44]

Nevertheless, Leakey's own position on the stone-tool argument was, to judge from his writings, highly ambiguous, not least because the species name he chose, *habilis*, seems to embody toolmaking as part of its definition, for it means "able, handy, mentally skilled, vigorous."

Faced with continuing criticism about the muddying of anatomical diagnoses by cultural inference, Leakey stated plainly at a meeting in Chicago in November 1965 that he separated himself from "any suggestion that you can use cultural evidence for any taxonomic [species diagnosis] purpose. The validity of *Homo habilis* rests entirely and solely on its morphological characters— not on any cultural ones."[45] And yet, not long prior to that he seemed to say quite the opposite at a gathering at the Cosmos Club in Washington, D.C.: "To me the most significant step that ever was taken in human history, the thing that turns animal into man, was this step of making tools to a set and regular pattern. This is why we chose that definition of *Homo*. . . . Once he had made the simplest of tools, he immediately opened himself a completely new food supply—and enhanced his chances of competing with other creatures."[46]

Once he had adduced the toolmaking association in the interpretation of *Zinjanthropus*, Leakey had made a logical snare for himself from which he never succeeded in breaking free. "Louis always insisted that he didn't take the tools into account in his interpretations," says his colleague Michael Day, an anatomist at St. Thomas's Hospital, London. "But it's clear that he always did."[47]

The principal sparring between Leakey and his critics over the naming of *habilis* took place on the pages of a now-defunct British magazine called *Discovery*. For a popular, nontechnical magazine, the exchanges were erudite, though pointed. Bernard Campbell launched the assault. Beginning in a truly British manner by first praising Leakey's efforts in discovering fossils, Campbell then says, "Their interpretation, however, is open to question." His principal complaint was, essentially, that Leakey did not understand the evolutionary process. When one species is transformed into another by natural selection, there will be a time—a transition period—when the living animals will look like neither the ancestor nor the descendant. Such a transitional form will have characteristics of both. "What we did not suppose was that the discoverer [of transitional forms] would create a new species to contain them,"[48] wrote Campbell. In company with other critics of the time, Campbell also took a verbal swipe at the use of cultural association in interpreting fossil hominids. "Dr. Leakey and his colleagues are not alchemists," he concluded, "and they must not expect us to accept their pronouncements in awed silence, unless they effectively demonstrate them to be justified."

Now, it is pertinent to note that at this point Campbell had recently completed a revision of hominid nomenclature in the literature. Like the Miocene apes that Elwyn Simons and David Pilbeam rescued from a chaos of inappropriate names, the hominids too had suffered from the excesses of super-splitters. There were more than 100 hominid taxonomic names in the literature, which Campbell had rigorously reduced to a small handful. "When *Homo habilis* came along, my outlook was to lump things together rather than split them,"[49] he recalls. "At the time, I could see no justification for a new species, and so I said it had to be either *Australopithecus africanus* or *Homo erectus*." Like the majority of the paleoanthropological community, Campbell has changed his mind now, and accepts *Homo habilis* as a valid species. "It was partly the later discovery of better specimens, particularly by Richard and his colleagues at Lake Turkana. But it also had to do with a change of attitude. I'm more of a splitter now than I was then."

Leakey was not especially wounded by the attack by Campbell, whom he regarded as a young upstart and a bookish person. Indeed, it served to strengthen his contempt for out-of-touch academics. "An unfortunate tendency has developed of late," he observed shortly after this episode, "for anthropologists who are mainly engaged in university teaching, rather than in actual field studies, to start lengthy discussion and criticism on the basis of preliminary reports, often without even viewing the original specimens, or casts thereof. This sort of controversy, often accompanied by dogmatic pronouncements, must be deplored."[50] This disdain of academia, however, did not prevent Leakey from striving fervently for the establishment's most sought-after accolade: namely, fellowship of the Royal Society of London. Although he had the support of Le Gros Clark and Sir Julian Huxley, there were others even more powerful who were opposed to Leakey. He never was elected.

What did wound Leakey over the *Homo habilis* affair, however, was the strictures from his onetime colleague Le Gros Clark. "I feel impelled to join with Dr. Campbell in his trenchant criticisms," wrote Le Gros Clark in July 1964. "Palaeontologists have become accustomed to the unfortunate proclivity of some field workers for coining names for fossils that later prove not to deserve them," he continued, corroborating Campbell's lumping approach. The unthinking handing out of new names was bad enough, he said, "But 'Homo habilis' is in a different category, for this nomenclatural twist has been made the occasion for claims that previous conceptions of hominid evolution need to be completely abandoned."[51] The main thrust of Le Gros Clark's argument was that the anatomical description given to "Homo habilis" (as he insisted on writing the name) was just as easily accommodated within *Australopithecus africanus.* "One is led to hope that ["Homo habilis"] will disappear as rapidly as he came," he concluded. "It certainly does not appear to merit prolonged controversy."

Leakey responded in the following month's issue. "I must confess," he began, "that I am somewhat surprised that Sir Wilfred Le Gros Clark, who has not yet had an opportunity of making a detailed study of *Homo habilis,* should nevertheless find himself able to state categorically [that the fossils belong to *Australopithecus*]."[52] Leakey then went on in great detail to explain exactly how the *Homo habilis* fossils differed from *Australopithecus.* He repeated the argument that he had advanced in his 15 November 1960 letter to Le Gros Clark, specifically that the coexistence of two hominids should not be a surprise given similar situations in the case of other animals. He also strongly rejected Le Gros Clark's suggestion that the naming had been done hurriedly and without due consideration. "Finally," he trumpeted, "I must protest against the misuse of quotation marks around *Homo habilis* . . . in Sir Wilfred's letter, as well as the failure to place them in italics. All these are valid names, in terms of the Zoological nomenclature, and are not nicknames. They should not be treated as such." It is interesting to see how the conventions of zoological nomenclature can be used so effectively to deliver not-so-subtle insults.

The exchange of letters continued, with other authorities joining in, including Phillip Tobias. In the final paragraph of his letter, Tobias states that he and Napier "are agreed that the new discoveries represent a stage in the development of man which neatly closes the gap between the most advanced *Australopithecus* and the lowliest *Homo*."[53] Tobias knew from previous private correspondence with Leakey that were he to state this position in public, Leakey would be forced to respond. And respond he did: "I believe that *Homo habilis* represents a distinct branch of *Homo* possibly leading to *Homo sapiens*, and do not see it as an intermediate link between *Australopithecus* and *Homo erectus*."[54] Leakey's statement was followed by *"This correspondence is now closed—Ed."*

This short exchange between Tobias and Leakey goes straight to the heart of what *Homo habilis* meant to Leakey. Like Sir Arthur Keith, he had always considered *Homo erectus* to be a specialized form that led nowhere. His view that the australopithecines were also too specialized, and in any case too late in the geological record, to be ancestral to *Homo* had still not been swayed by the majority opinion of the profession. Therefore, in his view, *Homo habilis* did not provide a neat link between these two forms, as Tobias opined, but instead served to sweep them both out of the way. *Homo habilis* was the direct precursor of modern man and the ancestral form that reached far back into the geological record. In short, it was the latest, but modified, manifestation of the ancient *sapiens* idea.

Leakey's reputation suffered through the *Homo habilis* controversy, principally because of his rapid switch of allegiance from Zinj. "I've often wondered what would have happened if we had found *Homo habilis* before Zinj,"[55] muses Mary Leakey. "I think Louis would have been in a much stronger position with his early *Homo*. His switch from Zinj to *Homo habilis* as toolmaker made people less than willing to take Louis seriously." *Zinjanthropus*, by the way, was relegated to *Australopithecus* by Leakey in the *Nature* paper that announced *Homo habilis*. Its anatomy had not altered, but Leakey's claim for its human status had. As it turns out, however, there do appear to be many intriguing parallels between the robust australopithecines and *Homo*, which modern researchers are now beginning to identify and puzzle over. What these parallels mean is as yet unclear.

The discovery, analysis, announcement, and subsequent debate of *Homo habilis* was really the last major event in Leakey's paleoanthropological career. He became more and more involved in fund-raising, for primate research as well as anthropology and archeology. But he was soon to watch the beginnings of his son

Richard's forays into the search for early man. In 1969, on his first major expedition to Lake Turkana (then called Lake Rudolf), Richard and his colleagues found a *Zinjanthropus*-like cranium and scraps of what looked like *Homo*, both of which were initially thought to be about 2.6 million years old. Louis told a meeting of the Leakey Foundation about it in October 1970: "Recently we have been able to trace the making of clearly defined stone tools back to 2.6 million years ago, and to find remains of both near-men and man associated in the same deposits at this remote period."[56] He then went on to reiterate what had been for him a guiding principle throughout his long career: "I honestly believe that within the next year or two, we shall carry the evidence of tool-making man . . . back to a much more remote period still, perhaps to 6 or 7 million years ago."

Shortly after this, the skull 1470 was found.

CHAPTER 8

THE LEAKEYS: SON

"It is *possible* that *Homo* goes back 5 million years. But this is only *a belief.* It is not based on the hard evidence,"[1] Donald Johanson told a large gathering at the American Museum of Natural History in New York in April 1984. "That belief is based on an idea that was originally started many years ago with Sir Arthur Keith, carried through by Louis Leakey, and is now held on to by Richard Leakey."

Richard Leakey's reaction to this suggestion is simple: "I really don't think there is a Leakey line on human origins, not since Louis died,"[2] he said recently. "Louis had strong views about human evolution. I don't."

During the past several years, daily newspapers and magazines have frequently featured reports and articles that appear to pit Leakey and Johanson against each other, a battle of young Turks. "RIVAL ANTHROPOLOGISTS DIVIDE ON 'PRE-HUMAN' FIND" and "BONES AND PRIMA DONNAS" are two examples of headlines that betray the tenor of the stories. It is certainly true that, as leaders in their discipline, they enjoy a public notoriety greater than even the most prominent nuclear physicist or molecular biologist. Leakey has dined at Ronald Reagan's White House and is to be seen promoting Rolex watches in full-page advertisements in *The New Yorker* magazine. "Some men merely make history. Mr.

Leakey redefines it," reads the copy line. Johanson, meanwhile, is an accomplished television personality and member of California's exclusive Bohemian Club, in company, for instance, with Henry Kissinger and Gordon Getty, and is director of his own internationally known institute. So the "prima donnas" epithet may not be entirely inaccurate.

But what of the profession's putative great intellectual divide? A recent origin of *Homo* versus an ancient origin? The Johanson line versus the Leakey line? How real is this? To what extent is Richard Leakey his father's intellectual son?

This chapter addresses these questions by tracing the story of Richard Leakey's initially reluctant entry into fossil hunting, the gradual transformation of his theories, and the eventual conflict with his friend and rival Don Johanson. It is a story that encompasses a period of unprecedented fossil discoveries in East Africa; a story that reveals the tenuousness of that evidence; a story that once again demonstrates what a very unusual science paleoanthropology really is.

Whether it is true or not that Richard Leakey inherited Louis' outlook on human origins, it is indisputable that he carried over a generous measure of the famous Leakey Luck. For at the age of twenty-four, on his very first full-scale expedition to Lake Turkana (then called Lake Rudolf) in 1969, he discovered two skulls of the sort for which his parents had to search patiently for more than thirty years. One of them, which is known by its catalogue number as KNM-ER (which stands for Kenya National Museums, East Rudolf) 406, was complete, intact, unbroken; it was clearly a Kenyan cousin of Zinj. The second, 407, was much more fragmentary and delicate; it was an enigma. Moreover, according to some hurriedly done radiometric dating of volcanic ash above the skulls, which was known as the KBS Tuff, they were both more than 2.6 million years old. To complete the largesse of this extraordinary field season, one of Leakey's colleagues, Kay Behrensmeyer, discovered pebble tools closely similar to the type that Mary Leakey had been familiar with for so long at Olduvai. The tools also appeared to be 2.6 million years old.

Until this point, Richard Leakey's involvement in and commitment to hominid hunting was minimal. A couple of years earlier he and a colleague, Kamoya Kimeu, had discovered a lower jaw that matched Zinj at Lake Natron, near the Kenya/Tanzania border. And in 1967 he had organized the Kenyan contingent in an international expedition to the Omo Valley in southern Ethiopia. But that, he says, was strictly for the fun. "I was legitimizing the use of other people's money to go to places I wanted to go and enjoy myself. I enjoyed running expeditions, and I was using my administrative know-how to run a scientific project." No one disputes that Leakey has organizational and administrative genius, a talent he displayed even at this young age. And, like his father, he has a very quick and attentive eye. It was while flying over the eastern shore of Lake Turkana on a brief visit to Nairobi during the 1967 Omo expedition that he noticed the potential of the vast deposits there. Perhaps, he speculated, they would contain fossils. A hurried inspection by helicopter confirmed that his eye had been true, and a preliminary expedition was swiftly planned for the following year, 1968. This vast stretch of fossil-rich sandstone sediments, which stretch some 40 kilometers from the lake's edge to the volcanic slopes of the Rudolf basin, is known generally as Koobi Fora.

That first excursion yielded a couple of scrappy hominid jaws, and a great deal of promise for the future. "At this point I still didn't have a job at the museum—that didn't start until October first, which was a week after the field project was due to end. And I had had no real involvement with fossil hominids, nor any intention of dealing with fossil hominids," insists Leakey. "I continued to nurture the notion that I would employ myself happily with fossil monkeys—for certain."

During the 1967 expedition to the Omo, Berkeley anthropologist Clark Howell had encouraged the young Leakey to pursue the study of fossil monkeys, and there was discussion of his perhaps doing a doctoral dissertation on them. But it was not to be, because then came the 1969 season. "With 406, the emotional charge I got from finding something myself, something everyone was going to like, was extremely powerful. I suppose it triggered off this paleontological possessiveness we all experience." From then on, Leakey was hooked. Thoughts of a four-or-five year academic slog with fossil monkeys evaporated. The 2,000 square kilometers of East Turkana deposits beckoned, just waiting to be explored, possibly rich in hominids.

But Leakey faced a problem. He had no formal education in anatomy or physical anthropology whatsoever, just years of experience in boiling down animal carcasses and reassembling the skeletons for sale to museums. What is more, his strong desire for independence from a young age had taken him away from his parents' activities. He was aware of what they were doing, but had no inclination to be involved or even take much interest, he says. His knowledge of fossil hominids was therefore rudimentary at best. "It didn't take a lot of intellectual power to see that 406 was the same as Zinj," Leakey recalls. "There was nothing difficult or controversial about that. But 407 was different. I talked a lot with Alan Walker about it. I also talked with Louis, and I was very influenced by him in this early stage."

That influence was obvious from the publications which resulted from this first field season, particularly in an article in *National Geographic* magazine, for so long Louis Leakey's mouthpiece to the world. Referring to the enigmatic 407 fossil, he wrote, "The face and jaws were missing, but the rest of the cranium bore few of the characteristics of an australopithecine. I felt, with mounting excitement, that the search for additional fragments and further study might show this to be no near-man, but perhaps even a species of the genus *Homo*."[3] Rhetorically asking, Who made the tools? he answered, "Not *Australopithecus boisei*, I felt." He went on to speculate that perhaps the toolmaker, 407, was a form of *Homo erectus*, much older than those at Olduvai. "Louis, right up to his death, considered that *Homo erectus* was a side branch, not ancestral to *Homo sapiens*," Richard now says. "It was in that context that we thought and I wrote about 407."[4]

It is worth noting, too, that Richard used the term "near-men" for the australopithecines, a term that was Louis Leakey's own and meant to express their similarity to, yet complete separateness from, *Homo*. Few other professionals employed the term, because most scholars considered that the australopithecines were in some way ancestral to *Homo*. Richard continued to use it for a few more years after this initial self-confessed aping of his father's views.

"In 1968, 1969 I had no preconceptions about human evolution, nor any desire to postulate a particular idea in terms of family trees,"[5] Richard now says. "I was very much under Louis' guidance. I didn't presume to have an opinion."

In addition to popular articles about the 1969 season, a clutch of four learned papers was dispatched to *Nature*, eventually to be published in the journal's 18 April 1970 issue. These comprised reports of the two skulls, by Richard; the stone tools, by Mary Leakey; the geology, by Kay Behrensmeyer; and the radiometric dating, by two British scientists, Jack Miller and Frank Fitch. In reporting on the oldest stone artifacts yet to be discovered, Mary Leakey stated that the discovery had simply confirmed her suspicions. "The multiple toolkit that was in use at Olduvai between 1.9 and 1.75 m.y. has led me to suggest," she wrote, "that systematic tool making must have been in practice for a considerable period before that time."[6] Richard Leakey echoed these sentiments in his *National Geographic* article: "These [Olduvai] implements, to my mind, show a degree of sophistication which implies that our ancestors began making tools well before [1.75 million years ago]."[7]

Here, then, right from the start, the large picture of East Turkana was drawn: *Australopithecus* and *Homo* were living side by side at a very early date; the two hominids had related but separate evolutionary histories; and *Homo* was the toolmaker. It was a picture that Louis Leakey would instantly recognize and be more than sympathetic with.

"From my point of view these discoveries . . . are particularly exciting," Louis Leakey told an audience at the Leakey Foundation in October 1969. "In 1933 I published on a small fragment of jaw we call *Homo kanamensis*, and I said categorically this is not a near-man or ape, this is a true member of the genus *Homo*. There were stone tools with it too. The age was somewhere around 2.5 to 3 million years. It was promptly put on the shelf by my colleagues, except for two of them. The rest said it must be placed in a 'suspense account.' Now, 36 years later, we have proved I was right, which to me is very, very, very satisfying."[8]

As things turned out, however, this initial picture would eventually crumble. Skull 407 was soon correctly identified as a female Zinj, not *Homo* at all, hence its lighter build. Geologists discovered that the two skulls, 406 and 407, had been incorrectly placed stratigraphically: they were in fact above the layer dated at 2.6 million years, not below it, and so were younger than this date. And in any case, that date itself would later be revised to 1.9 million. But all this was in the future, and the overview of the meaning of the East Turkana record was very much shaped by these first impressions.

As might be imagined, the revision of the 2.6-million-year date for the KBS Tuff to a much younger 1.9 million had a dramatic impact on the notion of an ancient *Homo*. A great deal of intellectual and emotional currency was invested in the earlier date, and so its drastic reduction was a long, confused and painful affair, stretching through to the mid 1970s. The story is chronicled in the following two chapters.

When Richard launched the program of exploration at Koobi Fora, he was faced with more than just getting scientific research under way. Namely, he had a reputation to establish. "When I took over in the museum in October 1968, one of the most barbed criticisms was that they called me administrative director, because I had no scientific qualifications, no respect, no clout in the scientific establishment,"[9] he now explains. "It struck me very strongly that this was something I wanted to change."

The way he planned to do it was by careful use of the scientific publications that would result from the analysis of the fossils discovered. The system of reporting the East Turkana finds to the scientific community was therefore established from the beginning. Richard would give a preliminary account in *Nature,* together with some interpretive observations; and then a more detailed description would follow in the *American Journal of Physical Anthropology,* written by Richard's scientific colleagues but usually including him as coauthor. "If we were getting all these results from Koobi Fora, I could initially present them in a scientific journal, I thought, without necessarily analyzing them in great detail, and without having a Ph.D.," reasoned Leakey. "The more serious work could then be done by people more qualified to do it. I decided to do the *Nature* papers, but initially declined to be involved with the *AJPA* papers. But my colleagues encouraged me to be involved and to be an author, so I did."

Although Leakey took advice and guidance on the *Nature* manuscripts, his wish to prove himself meant that they were very much his. Not surprisingly, therefore, the early attempts required a good deal of editorial and refereeing revision. And later on, the journal came to feel a little irritated, because its editors judged it was simply being used to catalogue each year's fossil haul. Nevertheless, a good professional relationship was established, and continues to this day, marred only by occasional outbursts on both sides. For Leakey, the strategy worked. By being a highly visible author in a most prestigious scientific journal, he gained the desired scientific respectability within the museum, even though, according to one colleague, the profession didn't really take what he wrote very seriously, in the early days at least.

Koobi Fora lived up to its promise, and year by year more hominid fossils—skulls, jaws, and limb bones—were recovered, providing the profession with a flow of new material it had never previously enjoyed. In his reports to *Nature* up to 1974 Leakey identified two types of hominids at Koobi Fora: these were male and female versions of Zinj, or *Australopithecus boisei*, as it is more properly known; and *Homo*, which he did not assign to any particular species. In 1972, for instance, he wrote, "The 1971 collection . . . confirms the contemporaneity of two hominid genera, *Homo* and *Australopithecus*."[10] He went on to argue that there was just one species of australopithecine at Koobi Fora, *Australopithecus boisei*, "within [which] there is considerable individual variation in size and I would consider that this can be explained in part by sexual dimorphism [males bigger than females]."

In common with other authorities, Leakey saw at Koobi Fora no evidence of the smaller australopithecine, *Australopithecus africanus*, which was generally considered by most observers, but not by Louis Leakey, to be ancestral to *Homo*. The implication of the Koobi Fora finds, as presented by Richard, was that there was simply too much uncertainty about *Australopithecus africanus* for it to be a good candidate as a precursor to *Homo*. And where it did appear to exist, it was clearly too recent in the record. "The concept of the gracile [delicate] australopithecine being ancestral to *Homo* in the Lower Pleistocene requires careful reexamination,"[11] he wrote in 1971. "The Rudolf material seems to confirm the view developed as a result of work at Olduvai . . . that *Homo* and *Australopithecus* are two quite separate and distinct early Pleistocene hominids." In other words, Richard was unambiguously supporting his father's position here.

The *Nature* reports were often accompanied by editorial comments in the journal's "News and Views" section, usually by an anonymous "Palaeoanthropology Correspondent." These comments are interesting in their reflection of prevailing opinion and reaction to the Koobi Fora finds. For instance, in the 2 June 1972 issue the correspondent wrote that "A tenet, which is basic to many theories of human evolution, is that no more than one type of hominid existed at any one time. This view has often been challenged, but the fossil evidence has been equivocal."[12] Given that this was some eight years after Louis Leakey and his colleagues announced the establishment of *Homo habilis*, it is clear how much doubt still surrounded the proposal. "Richard Leakey has very wisely refrained from allocating the [*Homo*] material to any particular species," the correspondent continued. "In this he

shows welcome, if almost unprecedented, restraint in a field clut-
tered with arbitrary and invalid nomina." Such as *Homo habilis,*
the reader is left in no doubt.

When, later that same year, in August 1972, the fossil-hunting
team at Koobi Fora began to uncover the fragments of the famous
1470 skull, the specter of *Homo habilis*—specifically, the bitter
controversy over the fossil that had engulfed Louis Leakey—was
thrust to the fore once again. Some of Richard's colleagues suggest
that once the initial excitement had subsided, he reacted as if he'd
seen the paleoanthropological equivalent of Banquo's ghost. But
the story is more complicated than this would imply.

The recovery of some three hundred fossil fragments and their
reconstruction into the remarkably complete skull 1470 took a
couple of months. But very quickly it had become obvious that
here was an unusually large-brained creature living (it was
thought) almost 3 million years ago. What was it? Not *Australopithecus,* for, among other things, its cranium was surely too big,
and it lacked the characteristic bony crest running along the midline of the head. What of *Homo habilis,* Louis Leakey's ancestral
candidate from Olduvai? This too was rejected. "The Olduvai material [of *Homo habilis*] is only known from deposits [younger
than] 1.96 m.y.,"[13] Leakey would later report in *Nature.* "At present therefore there does not seem to be any compelling reason for
attributing this species to the earlier, larger-brained, cranium from
East Rudolf." His conclusion was to "propose that the specimen
should be attributed to *Homo* sp. indet. [species indeterminate]
rather than remain in total suspense." In other words, having a
bigger brain than *Homo habilis*—800 c.c. compared with 680 c.c.
—and apparently being a million years older, 1470 was not considered to belong to the same species as its Olduvai cousin.

Leakey's course of action had been suggested by his British colleague Bernard Wood, of The Middlesex Hospital Medical School,
London. "As regards the name for the specimen . . . one wants to
be hyper-careful, because people will be just waiting for a repeat of
the 'launch' of *Homo habilis,* which wasn't entirely auspicious,"
wrote Wood on 15 October 1972. "I am sorry to be chicken about
this but the new material could be known by its number, and as
Homo sp."—which is precisely what happened. "The point I am
making is there is something very and unnecessarily emotive
about naming a new species. People will think that because you
give it a name you have all the answers—I can't truthfully say we

have all the answers. Neither do I want the obvious importance of the material obscured by prolonged wrangling in the press about nomenclature."

As might be imagined, there was a great deal of discussion around the Koobi Fora camp and back at the museum in Nairobi about what 1470 might be, for it was not unequivocally any known species. One point of uncertainty was the angle at which the face attached to the cranium. Alan Walker remembers an occasion when he, Michael Day, and Richard Leakey were studying the two sections of the skull. "You could hold the maxilla forward, and give it a long face, or you could tuck it in, making the face short,"[14] he recalls. "How you held it really depended on your preconceptions. It was very interesting watching what people did with it." Leakey remembers the incident too: "Yes. If you held it one way, it looked like one thing; if you held it another, it looked like something else. But there was never any doubt that it was different. The question was, was it sufficiently different from everything else to warrant being called something new?"[15]

They all agreed—Alan Walker, Michael Day, and Richard Leakey—that 1470 was quite separate from Zinj-like creatures. They all agreed too that there were some similarities with what had been called *Australopithecus africanus* from South Africa. But 1470's brain was much bigger—50 percent bigger, in fact. "This was when we first discussed whether it should be *Homo* or *Australopithecus*," recalls Richard. "Some clear differences of opinion eventually developed on this point."

Richard's view was soon formed, and firmly. "I think it's *Homo*, because of the big brain," he now says. "It represents an early part of the line in which the brain was expanding." Walker's assessment was different. "If you discount the large brain," he argued, "there are just too many similarities with *Australopithecus africanus* for them to be ignored."[16] These opposing opinions came into direct conflict in September 1973, during the final preparation of the detailed paper for the *American Journal of Physical Anthropology*.

Wood, Day, and Walker met Leakey in his office behind the museum and went over the final details of the descriptive paper, which was principally Walker's work. At the end of three hours of discussion on the content of the manuscript, Walker said, "Well, we agree on the paper, which you guys have just nit-picked and changed. What about the title?"[17] Leakey suggested "New Fossils from Koobi Fora of the Genus *Homo*." Wood would be happy with that. So would Day. Walker, however, most certainly would not. He reiterated his reasons why 1470 could legitimately be considered a large-brained *Australopithecus*, and made it clear he was strongly against being associated with the paper if there was any kind of statement linking the fossil with the genus *Homo*. Leakey persisted—at which point Walker got up, said, "Fine; in that case my name doesn't go on the paper," and left. "He just got up and walked out," recalls Day. "He didn't slam the door or anything. It wasn't melodramatic. But it was a very strong statement."[18] Leakey knew he must have Walker's name on the paper, as he had contributed so much to it. Eventually a compromise was reached. "We called the paper 'New Hominid Remains from Koobi Fora,'" recalls Richard. "I said I didn't think it was that important anyway."

The difference of opinion generated some tension for a while between the two men, particularly when Walker presented his reservations at an important scientific meeting in Nairobi that same September. "Leakey has said that 'There does not seem to be any basis for attribution (of this specimen) to *Australopithecus*,'" Walker told his audience. "But on the criteria given here it can be seen that several features that seem to be constantly found in *Australopithecus* are found in this cranium, not least of which is the relative proportion of the facial skeleton to the neurocranium. I am not suggesting that on this evidence 1470 is an *Australopithecus*, but I am arguing for caution, even when wanting to avoid keeping specimens in 'suspense account,' when we may be dealing with an extremely complex evolutionary problem."[19] Both Leakey and Walker now agree that their divided views illustrate cogently how very difficult it is to define what is actually meant by the genus *Homo*. Surprising as it may seem, there is still no good, generally accepted and crisp definition of *Homo*. And even if there were, the real evolutionary picture may have been so complex that some specimens from some time ranges inevitably would fall at the edges of such a definition.

The fact that Leakey went ahead and declared in *Nature* that 1470 was a member of the genus *Homo* provoked a variety of reactions from more distant members of the profession. Some suggested that he was reluctant to go to the logical conclusion and call it *Homo habilis* because of the furor that had surrounded its announcement and the odium that still attached to the name. And others said that it was for much the same reason that Leakey forbore to go to the logical conclusion and create a new species of *Homo*.

"I don't see how either could have been correct,"[20] Leakey now says, "because I was only dimly aware of the controversy surrounding *Homo habilis*. Yes, I remember Louis ranting and raving over Le Gros Clark and John Robinson. Mainly John Robinson. But I was never very interested in it, so I don't think it can have affected me very much." One reason no one was very keen formally to assign 1470 to *Homo habilis*, quite apart from the differences in brain size and supposed geological ages, was the disagreement that still surrounded Louis Leakey's species.

The major problem is this. When a new species is named, the author has to cite a so-called type specimen, against which other workers can compare similar fossil material. In addition, the author can add additional specimens, known as "paratypes" and "referred material," which allow further comparisons. With *Homo habilis* there are seven separate fossils cited in all. Now, according to many authorities, this array of fossils erroneously includes representatives of *two* species, not just one, as it is meant to. Some of these fossils are accepted as *Homo* while others may well be *Australopithecus africanus*. So, although as a collection these seven fossils are meant to define *Homo habilis*, instead they cast a veil of ambiguity over the species. This creates a problem for people who, when analyzing a new fossil, wish to know if it is *Homo habilis* or not. The answer is, well, it depends on what you mean by *Homo habilis*. The required comparison, formally speaking, cannot be unequivocal because of the mixed sample.

"To call 1470 *Homo habilis* would therefore demand that the whole species be formally revised," explains Leakey. "But by default, this is what it has come to be attributed to."

While these early discussions about the 1470 enigma were still simmering, another discovery was made at Koobi Fora that added further complexity to it all. Known by its catalogue number of KNM-ER 1813, the cranium was an immediate puzzle and the subject of strong, and continuing, disagreement. The teeth of 1813 were just like one of the Olduvai fossils that was supposed to be *Homo habilis*. And yet its cranium was tiny. Number 1813 simply couldn't be *Homo* with a brain like that was the immediate reaction, which observation threw further doubt on the integrity of the supposedly diagnostic *Homo habilis* collection from Olduvai. This time Leakey and Walker were in agreement, thinking the fossil might be *Australopithecus africanus,* the first to be found at Koobi Fora. Wood, however, disagreed, and he now thinks the skull might be a new species of *Homo.*

If 1813 were indeed *Australopithecus africanus,* the picture of human origins as portrayed by Leakey's interpretation of the Koobi Fora fossils would become very much clearer. In his annual report to *Nature,* Leakey argued that 1813 was one of the more delicate type of australopithecine. He noted the presence at the same period of its Zinj-like relative and of *Homo,* as represented by 1470. And drew a modest conclusion: "all these forms may be traced back well beyond the Plio-Pleistocene boundary."[21] The real implication of this statement was left to the anonymous paleoanthropology correspondent to articulate in "News and Views": "If one accepts that gracile australopithecines and *Homo* coexisted in the early Pleistocene, then one must also accept that [*Australopithecus africanus*] does not, indeed cannot, represent the ancestral hominid group. The gracile australopithecines are widely accepted as the basal hominid stock and yet Leakey's current classification of the early hominids implies that the known members of this group coexisted with the genus *Homo* and were in fact too late to provide the ancestral population."[22]

So if Leakey was correct, the line leading to modern humans did indeed have an ancient origin, and was separate from the australopithecines: these last were near-men, not ape-men, in Louis Leakey's terminology. In an article published shortly after the discovery of 1470 and 1813, Richard Leakey wrote the following: "I feel confident that one day we will be able to follow man's fossil trail at East Rudolf back as far as four million years. There, perhaps, we will find evidence of a common ancestor for *Australopithecus*—near-man—and the genus *Homo,* true man."[23] With the passage of the four very eventful years of exploration at Koobi Fora since that first spectacular season of 1969, the emerging interpretation of the fossil evidence was still very much one with which Louis Leakey would have had full sympathy.

But those first three seasons at Koobi Fora did more than bring Richard into intellectual sympathy—if not total agreement—with his father. Specifically, the discovery of 1470 did for Richard Leakey what Zinj had done for Louis: it catapulted him to worldwide fame and recognition. Shortly after the cranium had been assembled he took a cast of it to London, where he announced the discovery at a meeting of the Royal Zoological Society. Leakey took care to alert the press to the impending event beforehand, even though the organizers of the meeting, and particularly Lord Zuckerman, were not especially anxious to have newshounds drooling over significant new skulls from Africa. The meeting had been arranged as a centenary celebration of the birth of Sir Grafton Elliot Smith, Zuckerman's mentor, and it was not considered appropriate that attention be diverted from the central purpose. Concern was particularly keen on this point, because just a few weeks before the meeting a newly published book, *The Piltdown Men*, by Ronald Millar, had fingered Elliot Smith as the perpetrator of the hoax. Both Zuckerman and his colleague Joseph Weiner, who had participated in the exposure of the Piltdown hoax, angrily rebutted Millar's charge.

The upshot of the occasion was that hurried rearrangements had to be made, mainly by Bernard Wood, so that the press could meet Leakey safely distant from the meeting rooms of the Zoological Society. Zuckerman had forbidden such a gathering at the Society. The 1470 find and Richard Leakey were on the front pages of major newspapers in Africa, Europe, and the United States the following day. A new Leakey era had begun.

Part of becoming accustomed to this new era, for Leakey, was accepting that some in the field would always regard him as an amateur, because of his lack of academic qualification in paleoanthropology. Zuckerman explicitly did so at the Zoological Society gathering. "Mr. Chairman, may I first congratulate Mr. Leakey, an amateur and not a specialist, for the very modest and moderate way he has given his presentation,"[24] he said, apparently unaware of lifted eyebrows his remarks occasioned among his audience. "May I also express my personal gratitude and certainly the gratitude of many others who have worked with him and his father for the work they have done, not as anatomists, as Mr. Leakey pointed out, not as geochemists or anything else, but just as people interested in collecting fossils on which specialists can work."

One of the ironies of this occasion, which is usually little mentioned, was the surprising haste with which Zuckerman was prepared to accept Leakey's presentation. His Lordship's scorn for the level of competence he sees displayed by paleoanthropologists is legendary, exceeded only by the force of his dismissal of the australopithecines as having anything at all to do with human evolution. "They are just bloody apes," he is reputed to have observed on examining the australopithecine remains in South Africa.

Since his emigration to England from South Africa in 1926, Zuckerman had become extremely powerful in British science, being an adviser to the government up to the highest level. During the 1940s and '50s, however, while at Oxford and then Birmingham universities, he had vigorously pursued a metrical and statistical approach to studying the anatomy of fossil hominids. No secure inference could be drawn without such an analytical approach, he urged, and it was on this basis that he underpinned his lifelong rejection of the australopithecines as human ancestors. His reception of 1470, however, was different.

"[Had] today's discovery been reported in this Society when the *Australopithecus* skull first rested on the speaker's bench in our old meeting room, any amount of time would have been saved,"[25] he observed after Leakey's presentation. "People would not have been turning themselves inside out . . . in order to establish anatomical conclusions which were nonsensical. You may not have intended to, but you have demolished all that with your skull." To which Leakey replied, "I am quite pleased I have."

Later, during a lecture at the California Institute of Technology, in Pasadena, Zuckerman said, "[Leakey's 1470 skull] shunted the australopithecines to the sidelines, where they have always belonged."[26] About Richard Leakey's interpretation of 1470 as *Homo*, Zuckerman said, "I accept his statement, although Mr. Leakey is no anatomist." Zuckerman apparently required no metrical analysis to see that Leakey's new fossil supported his view of human origins.

Leakey had certainly been looking for attention with his announcement of 1470 at the Zoological Society meeting, but was surprised at the degree of it. "Having got it, it was a question of, What next?"[27] he recalls. "We clearly needed a lot more fossils to address the questions that everyone was interested in."

Koobi Fora continued to yield hominids with each year's field season, but doubt over the accuracy of dating of the infamous 2.6-million-year KBS Tuff threw a growing cloud of confusion over interpretation of the material. Leakey, meanwhile, was following in his father's footsteps by making frequent trips to the United States for fund-raising and lecture tours. Louis was Richard's mentor in this respect: "I greatly admired his ability to inspire people.

I try to pattern what I do in public on this, for I saw so many go so far on his words." By all accounts, Richard is at least as accomplished in this direction as his father, if not more so. He frequently speaks to audiences of thousands, who eagerly pay to hear the continuing Leakey tradition. Richard also established his own fund-raising organization, the Foundation for Research into the Origins of Man, FROM, which was based in New York. Like the Leakey Foundation, which had been established for Louis and from which Richard remained aloof, FROM raised money for distribution to researchers in paleoanthropology, archeology, and primatology. Donald Johanson was one of its board members, until he resigned to set up his own institute at the end of 1980.

Leakey's fame grew, and in November 1977, shortly after the publications of his best-selling book *Origins*, he was featured on the front cover of *Time* magazine, together with a somewhat grotesque mock-up of 1470. That issue was one of the all-time biggest sellers—"bigger even than the one with Cheryl Tiegs," Leakey notes with more than a little amusement. Later, during 1979 and 1980, he produced seven one-hour programs with the BBC on the whole sweep of human prehistory, not just the early stages with which he is usually identified. The series was called *The Making of Mankind*—but because of the culmination of chronic kidney failure, it was very nearly the undoing of Richard Leakey. An organ transplant from his younger brother Philip midway through the filming saved his life.

Leakey had long planned to do such a television series and, in fact, had envisaged it as the end to his involvement in human-origins research. "I had seen it as a conclusion to a career," he now says. "Most people are unaware of this, but I don't and never have viewed the pursuit of fossils as my only career. The work at Koobi Fora has been a major part of what I've done, that's true, and I've greatly enjoyed the fieldwork, the outdoor life, the organization and the notoriety. I've appreciated the fact that the notoriety has allowed me to do things I might otherwise have been unable to do, like build this splendid museum here in Nairobi. But I also enjoy and am committed to many other things, including education through the many different departments at the museum, and wildlife conservation. And I'm proud of having built up Louis' primate center to be a place for international scientific research."[28]

As it happened, the filming and the hospitalization were going on against a rising background of personal abrasion with Johanson and his colleague Tim White, which enhanced Leakey's inclination to step out of paleoanthropology for good. But his colleagues —Alan Walker, the late Glynn Isaac, and David Pilbeam—prevailed upon him to abandon that idea. "Alan said, 'Why give up something you enjoy so much, the fieldwork?' Glynn said, 'It's never bothered your family before to be the brunt of these kinds of things; come on, straighten your back and continue.' And David persuaded me that I really did have a contribution to make. So I stayed with it, and I'm enjoying myself again."

The new fossils that Leakey felt were needed to take the story of human origins beyond what 1470 and 1813 indicated began to be discovered during the mid 1970s, but not from Koobi Fora. Mary Leakey had returned to Laetoli, a site some 40 kilometers southwest of Olduvai that Louis had visited in the 1930s. In addition to the dramatic footprint trails that Leakey and her colleagues uncovered there, she found fragments of some dozen or so very primitive-looking hominids, mainly parts of jaws and teeth. Both the footprint trail and the hominids were dated to an incredible 3.75 million years. At about the same time, Johanson and his French colleagues Maurice Taieb and Yves Coppens were beginning to discover the very rich hominid-fossil deposits in the Afar region of Ethiopia. During 1974 and 1975, Lucy and the "First Family" were recovered, and paleoanthropology would never be the same again.

The Laetoli and Afar fossils now formed the focus of new ideas in human-origins research, and this all depended, of course, on the interpretation of what, exactly, they were. The full story of their formal entry into the scientific literature is told in two later chapters. The fossils would become the focus of a growing intellectual divide between Leakey and his friend Don Johanson, a divide that is seen by some professionals as a perpetuation of "The Leakey Line" on human origins versus Johanson's newer interpretation. It is an ongoing manifestation of the debate over an ancient versus a recent origin for *Homo*.

Mary Leakey had a clear idea what her Laetoli fossils were: "They have many characteristics that are similar to those of *Australopithecus*," she observed, "but I consider that they are the only possible candidate for an ancestral form of *Homo* at this particular date."[29] Richard noted that "Not all prehistorians would agree with this position, but I feel that a good case can be made out for its support." These judgments were based only on jaw fragments, which at best are a limited source of diagnostic information—a fact that Richard Leakey acknowledges. "It is very unlikely that mandibular or dental morphology alone will be sufficient for positive identification,"[30] he said in 1975.

The much larger sample from Johanson's site, which displayed a wide range of individual variation in size, was more of a diagnostic challenge. Mary and Richard Leaky visited the site briefly during the 1974 season and formed the impression that some of the larger specimens were, like Laetoli, *Homo*. Johanson at the time concurred, and said as much in the March 1976 issue of *Nature*. The smaller specimens, he suggested, might be related to *Australopithecus africanus*. And there was even a possibility that some of the fossils might be members of the large australopithecine species, *Australopithecus boisei*. The age of Johanson's fossils was considered to be somewhere between 3 and 3.5 million years.

The Laetoli and Hadar fossils in combination had a tremendous impact on the profession. Hominids had never before been recovered from this age range, and their quantity gave reasonable hope of a secure diagnosis. "Such evidence indicates that the earlier and simpler models of 'straight-line' evolution (for example, *Ramapithecus—Australopithecus—Homo*) could not be substantiated in the fossil record,"[31] an anonymous correspondent wrote in *Nature* in December 1976. "Thus the evolution of the Hominidae is more complex, and ultimately more interesting, than previous indications." The correspondent then went on to identify a key implication of the new evidence: "the genus *Homo* may be far older than previously thought." This would have been music to Louis Leakey's ears. It certainly pleased Richard.

As it happened, these developments were taking place at a time when it was becoming inescapably clear that the originally proposed date of 2.6 million years for 1470 was incorrect, and 1.9 million was more likely to be right. Support for an ancient origin of *Homo* from the Koobi Fora fossils was therefore crumbling. Nevertheless, Leakey was able to adduce these new finds from the Hadar and Laetoli as corroboration of the old idea. Picking up on the theme as expressed by the anonymous correspondent in *Nature*, he said in the same year: "I am reluctant to anticipate further new discoveries, but I would expect that the genus *Homo* will eventually be traced into the Pliocene at an age of between 4 and 6 million years, together with *Australopithecus*."[32] Leakey's view has remained more or less the same through to the present.

During 1977 and 1978, Johanson, collaborating with Tim White, began to change his opinion about the Hadar fossils. Instead of representing two or even three species, the Hadar fossils, Johanson came to believe, were from just one species that displayed a great deal of individual variation in size. Moreover, he and White identified the Laetoli specimens as belonging to the same species, despite their geographical distance of some 1,500 kilometers and half a million years' separation in time. Midway through 1978, Johanson and White, in company with Yves Coppens, formally gave the appellation *Australopithecus afarensis* to the fossils, making it the first major new hominid species to be named in 15 years. The previous one, of course, was *Homo habilis.*

Having named the fossils in the dry, objective tones required by international zoological convention, Johanson and White then went ahead and explained how the new species might affect interpretation of the human family tree. "Because of their great age, abundance, state of preservation, and distinctive morphology, the Laetoli and Hadar fossils provide a new perspective on human phylogeny during Pliocene and Pleistocene times,"[33] they wrote in the 26 January 1979 issue of *Science*. That perspective was seductive in its simplicity. *Australopithecus afarensis*, they proposed, was the only hominid living in the 3-to-4-million-years ago period, and was ancestral to all later hominids. By this account, the origin of the *Homo* line was therefore somewhere between 3 and 2 million years ago—which is in stark opposition to the Leakey view of the world.

This conception of human origins gives a very simple shape to the family tree: a neat Y-shape. The vertical stem of the Y is *Australopithecus afarensis*, which branches one way to produce *Homo habilis*, leading eventually to *Homo sapiens*, and branches the other way to produce *Australopithecus africanus*, which leads to its more robust cousin, *Australopithecus boisei*, and then to extinction.

"Immediately after the announcement of *Australopithecus afarensis* there was an instant reaction from Richard Leakey,"[34] recalls Johanson. The reason? "Here we had placed *Australopithecus*, the ape-man, in our direct ancestry. We had suggested that *Homo* could be traced no further back than 2 million years." The two men, he said, were intellectually divided on their view of human origins: Leakey versus Johanson, ancient *Homo* versus recent *Homo*.

Less than a month after the publication of the *Science* article, a photograph of Leakey and Johanson appeared on the front page of *The New York Times*, showing the two anthropologists apparently locked in disagreement. "Richard Leakey, the Kenyan anthropologist, is challenging the announcement made last month by two American scientists that they had discovered a new species,"[35] ran the copy. "Although honest differences of opinion are common enough in all sciences, there are overtones of a confrontation between the two anthropologists, each of whom leads a major fossil-hunting expedition in eastern Africa. The two men have often been viewed as rivals."

The occasion was a series of lectures in Pittsburgh held by Leakey's organization FROM. Leakey forbore to challenge *afarensis* during his formal presentation, but was drawn into discussion by reporters after the session was over. "I think Don was right the first time," said Leakey, referring to the initial publication on the Hadar hominids in *Nature* in 1976. "They're sampling different populations, *Homo* and *Australopithecus*." Johanson firmly disagreed. He adduced the First Family in support of his position, pointing out that the full range of variation of size and anatomy in the whole Hadar collection was to be seen in this group of thirteen individuals who had all been killed together in some kind of catastrophe, possibly a flash flood. They were probably close relatives who lived and foraged together, much as chimpanzees and baboons do today. The range of variation in size and anatomy was therefore simply an aspect of one population of one species, argued Johanson, not evidence of the mixing of two or more species. Subsequently it turned out that the geology does not support the idea of an instant catastrophe, so it appears that the First Family may not be a group of related individuals after all. More likely, their skeletons had become buried and fossilized separately over a long period of time, thus giving no indication of whether they belong to one species or more.

Leakey also said that he had fossil evidence which would show Johanson to be wrong in his proposal. He could not go into detail, he said, because the fossils had not yet been published. "The material I've got is very insignificant, but there's enough of it to challenge Don with," he argued. "It gives me the right to offer my opinion." These fossils, which have only recently been published, are eight teeth—four molars and four premolars—collected in 1978 from the south end of the Koobi Fora region and dated at about 3 million years. According to Alan Walker, the teeth are "dead ringers" for the kind found at the Makapansgaat cave in South Africa, which would make them *Australopithecus africanus*. If this is what they are, and they are 3 million years old, then *Australopithecus afarensis* cannot be ancestral to all later hominids because it was already contemporary with one of them.

"Yes, I know those teeth,"[36] says Johanson. "I was in Nairobi for a FROM meeting in November 1978. Richard showed them to me and said, 'Here, what do you think of these?' I said something like 'Well, they look like the teeth from Makapan, but they also look like *afarensis*.' To which Richard replied, 'Oh, you disagree with Tim, then?'" White had been shown the teeth some months earlier. "I had been given the 'nose-pass job' in front of Walker and Pilbeam in the hominid room at the museum,"[37] recalls White. "I said I thought they were *afarensis*." Yves Coppens, the third namer of *afarensis*, also saw the teeth. In his opinion, they were like the species from Makapan, *Australopithecus africanus*. All of which indicates, perhaps, that teeth are not necessarily very diagnostic, especially from two closely related species—if indeed there are two species there to be differentiated.

The previously cordial relationship between Leakey and Johanson began to spiral downward following the Pittsburgh FROM conference. The reasons are many, some of which involve serious personal accusations. Ostensibly, however the rift was supposed to be the result of the professional disagreement. "There is a controversy that has been going on now for nearly three years between Richard and myself, and it specifically focuses on the family tree,"[38] said Johanson on the May 1981 *Cronkite's Universe* program. But in his book *Lucy*, which was published in 1981, Johanson was sharply critical of Leakey's professional conduct, particularly in his handling of the revision of the 2.6-million-year date for the KBS Tuff (the subject of the following two chapters). By this time Johanson had resigned from the FROM board after some keen differences of opinion with its chairman, Leakey.

In 1981 Johanson left the Cleveland Museum, where he had worked since gaining his doctorate, and established his own research center, the Institute of Human Origins, in Berkeley. By now the finder of Lucy was clearly the new king of anthropology in the eyes of the U.S. print and television media. But just as everything appeared to be developing so well, catastrophe occurred.

Late in 1982, the Ethiopian government halted all prehistory research involving foreign scientists. The stated reason was that the relevant ministry needed time to draw up guidelines for such research, at least in part to ensure the proper involvement of local scientists. In fact, the ban came down amid a welter of accusations and rumors of the most florid nature, involving the alleged theft of fossils, CIA connections, and bribes. Perhaps it is not surprising, therefore, that instead of a mere twelve months for resolution, the ban remained in force through 1986 (and continues at the time of writing). As might be imagined, this has been a source of immense professional frustration to Johanson and his newly founded institute.

At one point the Ethiopian commission reviewing the antiquities policy consulted Leakey, as head of the National Museums of Kenya, seeking information on how his country organized this kind of research. Relations between Leakey and Johanson had deteriorated so much by this time that Johanson's suspicions were stirred by this contact. "From what I understand from certain sources, Richard has been undermining our efforts in Ethiopia."[39] Johanson said recently. "I don't have any documents to show you, copies of letters or anything, but that is what I understand." There is indeed no evidence that Leakey acted with the Ethiopians in any way other than as a fellow Third World administrator experienced in antiquities policy.

Johanson's closest colleague Tim White has been central to the growing disaffection between Leakey and Johanson. White, who is widely acknowledged as one of the profession's ablest morphologists, worked closely first with Richard Leakey at Koobi Fora and then with Mary Leakey at Laetoli. The deep affection and loyalty that White felt for the Leakeys in the mid 1970s has since been transformed into an equally passionate animosity. "What the Leakeys say is that we had a falling-out over *afarensis*,"[40] White now bitterly charges. "We didn't have a falling-out over *afarensis*. We had a falling-out over who dictates research results. We had a falling-out over who gets all the credit." The two men exchanged increasingly bitter correspondence during the early 1980s, in which White pursued and expanded these charges and Leakey demanded a public apology for what he considered were baseless accusations publicly flaunted. The personal and professional relationship between Leakey and White effectively ceased in 1985.

The effects of the feud are several. One is that in a field where opinion can sometimes overwhelm irrefutable fact, the natural tendency to polarize to one viewpoint against another has been enhanced. There are, of course, genuine differences of professional opinion among scholars as to the interpretation of the Hadar and Laetoli fossils, and the field remains roughly equally divided over whether there is just one species there or more. But these opinions are sometimes worn more like battle colors rather than scholarly points of view.

A second effect is that there is an even greater emphasis than usual on personality. "In Kenya, Richard Leakey is the guy who determines who gets access to what sites,"[41] observes William Kimbel, with none-too-subtle implications. Kimbel, who took over from Johanson at the Cleveland Museum, is now president of Johanson's institute in Berkeley. "You can get attention by throwing a tantrum," says Russell Tuttle, a paleoanthropologist at the University of Chicago. "The problem is, Johanson wants to hog the limelight." And so it goes.

"One of the sadnesses I have is that so often when the general public hears about studies of human origins, they hear it in the context of emotional arguments, personality cults, and personality assassination attempts,"[42] Leakey has remarked. "I think these studies are more important than that."

"This *afarensis* affair has caused so much bitterness,"[43] Leakey said recently. "It is all extremely unfortunate." He thinks, however, that the press has made too much of the affair. "Certainly far too much," he told an interviewer for *Omni* magazine in 1983. "But I'm not sure the press has been unaided. I certainly have always refused to talk about this issue to the press. I have never studied the Hadar material. I've simply expressed the professional view that Johanson has one explanation of its significance, but that there are other explanations. I've always felt that the material simply didn't warrant a dogmatic position."[44] David Pilbeam agrees. "I've always thought there was a possibility that there was more than one species,"[45] he said recently, "but there's not enough material to make a solid case either way."

Richard Leakey began his paleoanthropological career using the term "near-men" for the australopithecines—a clear emblem of "The Leakey Line." He no longer employs this appellation, but, it's fair to say, he nevertheless still holds close to the vision of human origins that was Louis'. Father and son would disagree on one aspect of the picture: Louis saw *Homo erectus* as a side branch to the main stem; Richard does not. But on the position of the australopithecines, they are more or less inseparable, even if their modes of expression of it might appear to differ in cogency. Louis dogmatically pushed *Australopithecus* to one side. Richard is more circumspect. "Somewhere down the line there must have been something that was ancestral to both *Australopithecus* and *Homo*,"[46] he now says. "Whether you call that *Australopithecus*, it doesn't matter. The point is, I don't think it's been found yet."

Richard's current assessment is best expressed by the following: "I envisage two species of hominid living at Hadar just over three million years ago: a larger species which was a primitive form of *Homo*, and a group of smaller hominids belonging to a previously unknown *Australopithecus* species [*afarensis*]. The *Homo* line had to arise at some point in time, of course, but I suspect that the instant is further back in time than the Hadar and Laetoli deposits. Given the well developed nature of *Homo habilis* at around two million years ago, and what I see as the diversity of hominids at three-and-a-half to three million years ago, it seems a fair guess that the *Homo* line may have evolved initially as long as five million years ago."[47]

It may be true that Richard Leakey has so deeply, yet unconsciously, imbibed Louis Leakey's outlook on human origins that willy-nilly, he unswervingly interprets the evidence through his father's eyes. Equally, it may be true that what Richard Leakey sees in the evidence is what is truly there to be seen. Louis Leakey may have been mainly right. But as yet, no one knows.

Ian Tattersall (center) supervises the display of some of the 40 or so original hominid fossils that formed the American Museum's *Ancestors* exhibition in April 1984. It was an emotional event for participants in a specially arranged symposium, and someone said, "It was like discussing theology in a cathedral."   © *American Museum of Natural History*

Misia Landau realized during her studies of anthropological literature that many accounts of human origins were presented like storytelling. "I knew I had made a discovery," she now says. "It was like finding a fossil." © *Ali Farhoodi*

Henry Fairfield Osborn, president from 1908 to 1935 of the American Museum of Natural History, saw the rise of civilized humanity as an evolutionary prize won in the face of hardship. No struggle, no prize. Photographed here in 1933, he believed that "the rise of man is arrested or retrogressive in every region where the natural food supply is abundant and accessible without effort." © *American Museum of Natural History*

Grafton Elliot Smith, a leading British anthropologist of the early twentieth century, used his characteristically florid prose to tell tales of the drama of human origins. He writes of "... the wonderful story of Man's journeyings towards his ultimate goal," and "... Man's ceaseless struggle to achieve his destiny." © *University of London*

Sherwood Washburn was one of the first scientists outside Yale to read Landau's thesis. "I quickly became fascinated by it. It is a very useful idea. It makes it much easier to change your ideas. Once you have, in quotes, a Scientific Theory, all capital letters, there is a great resistance to change." © *University of California, Berkeley*

Robert Broom, seen here at Sterkfontein quarry, was responsible for some of the earliest hominid fossil finds in Africa. He was also the most extreme among his contemporaries in seeing humans as the goal of evolution. "Much of evolution looks as if it had been planned to result in man, and in other animals and plants to make the world a suitable place for him to dwell in." © *British Museum (Natural History)*

David Pilbeam, who was Misia Landau's thesis adviser at Yale, says that "our theories have often said more about the theorists than they have about what actually happened."
P. Kain © *Sherma*

Niles Eldredge (right) and Ian Tattersall, paleontologists at the American Museum of Natural History, said in a book titled *The Myths of Human Evolution* that "science *is* storytelling, albeit of a very special kind."
© *American Museum of Natural History*

Raymond Dart with the Taung skull
in February 1925, just after the
announcement in *Nature*. For Dart,
the clue to the Taung child's family
connection was in its brain. "That's
what put me on to the idea that the
fossil wasn't just an ape's," he now
says. "Without that endocast, and
without my experience in neurology,
I doubt I would have thought it was
a hominid." © *Barlow/Rand*

Phillip Tobias, who now occupies
Dart's professorial chair at the
University of the Witwatersrand in
Johannesburg, shows the Taung child
fossil to scientists at the Taung
Diamond Jubilee Symposium.
© *R. Lewin*

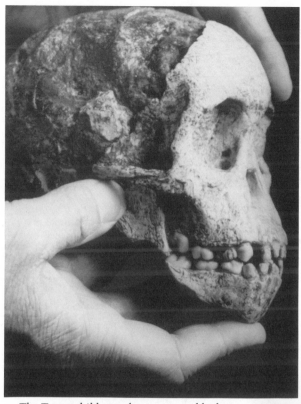

The Taung child was about 3 years old when he died, as revealed by the stage of development of the milk teeth. Part of the braincase of the fossil was never recovered, but the stone cast of the child's brain is remarkably intact. It was the shape of the brain, particularly at the back, that gave Dart a clue that this was not just another ape.
P. Kain   © *Sherma*

Sir Arthur Keith was one of the leading British anthropologists at the time of the Taung child discovery, but the fossil failed to impress him. "It may be that *Australopithecus* does turn out to be 'intermediate between living anthropoids and man,' but on the evidence now produced one is inclined to place *Australopithecus* in the same group or subfamily as the chimpanzee and gorilla," he wrote in the journal *Nature*.   © *Royal College of Surgeons*

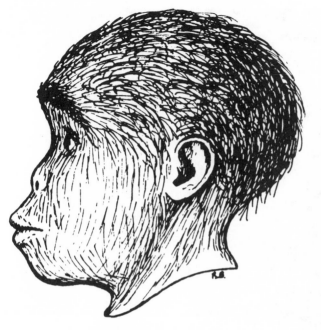

"Head of the little boy from Taung." This picture was drawn by Robert Broom, who was skilled at sketching reconstructions of fossil creatures from the smallest of fossil evidence. © *British Museum (Natural History)*

"Head of the young man from Kromdrai." Robert Broom's sketch of the robust form of *Australopithecus.* © *British Museum (Natural History)*

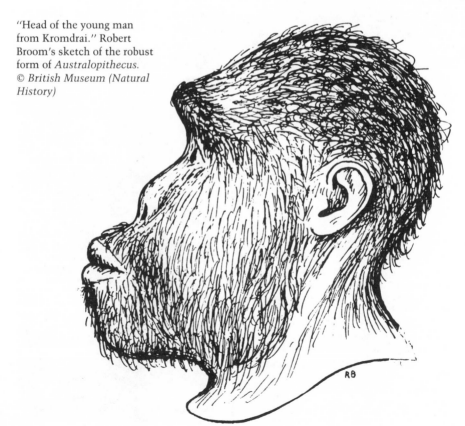

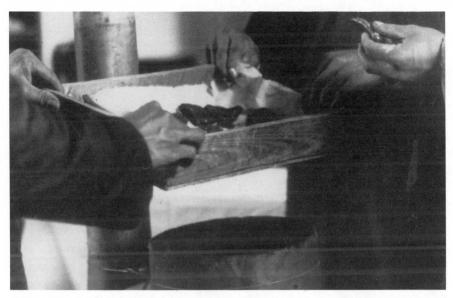

Peking Man fossils being dried immediately after excavation in 1929 (above) and site of further excavations a decade later (below). These fossils from China were quickly accepted as members of the human family, because they fitted people's preconceptions. "There was a climate of opinion that favored discoveries made in Asia but not the 'silly notion' of small-brained bipeds from Africa," said Sherwood Washburn recently.   © *Institute of Paleontology, Peking*

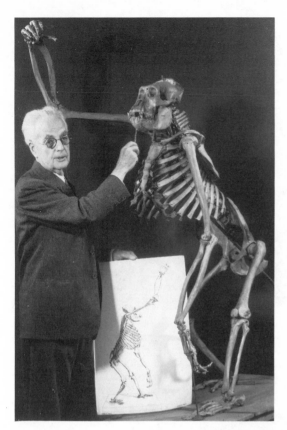

William King Gregory, Osborn's close colleague but intellectual rival at the American Museum, taken in 1951. "I fear we have reached opposite conclusions," Gregory wrote to Osborn in November 1920. " 'Back to Huxley and Darwin' is the motto of my conclusions." © *American Museum of Natural History*

Neanderthal man, a 50,000-year-old specimen from La Ferrassie in France. One of the most striking features of Neanderthal skulls is the way the face projects forward, as if the nose had been pulled out, taking the rest of the face with it. The bone is thicker than in modern humans, and there is a prominent ridge above the eyes. Although the brains of Neanderthals were often bigger than those of modern humans, most anthropologists of Marcellin Boule and Arthur Keith's time considered them to be of poorer quality. © *Margo Crabtree*

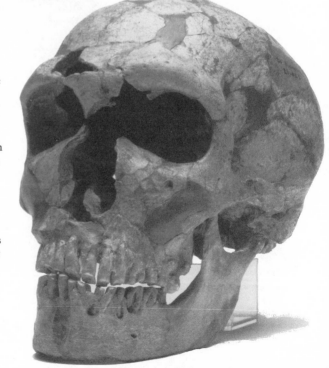

Depictions of Neanderthal (top) and Cro-Magnon (bottom), drawn in 1915 by C. R. Knight under the direction of Henry Fairfield Osborn. The drawings reflect not just a strict assessment of differences in anatomy but also supposed differences in bearing, nobility, and civilization. © *American Museum of Natural History, drawn by C. R. Knight*

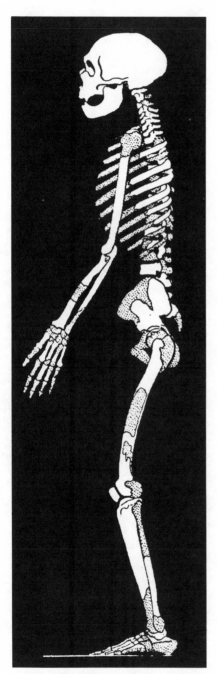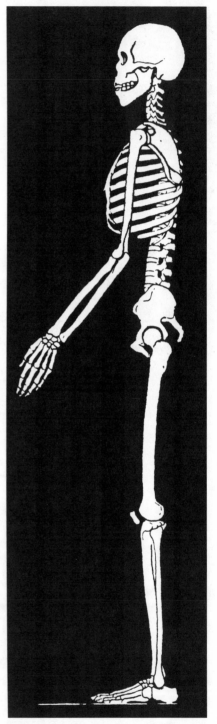

Neanderthal (left) and *Homo sapiens*, as depicted by Marcellin Boule in 1921 in his book *Fossil Men*. Note the bent-knee, stooped posture attributed (incorrectly) to the Neanderthal skeleton by Boule.

A discussion of the Piltdown skull. Back row, left to right: F. O. Barlow, Grafton Elliot Smith, Charles Dawson, and Arthur Smith Woodward. Front row, left to right: A. S. Underwood, Arthur Keith (examining the skull), W. P. Pycraft, and Ray Lankester. The Piltdown man "fossil" fitted most British anthropologists' preconceptions so closely that, observed Sir Wilfred Le Gros Clark, "none of the experts concerned were led to examine their own evidence as critically as otherwise they would have done." Taken from the portrait painted by John Cooke, *R.A.*, in 1915. *© British Museum (Natural History)*

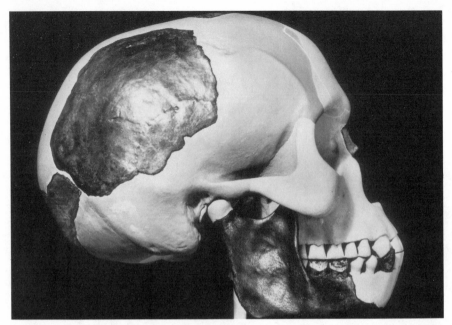

The Piltdown skull reconstructed. The very humanlike form of the cranium and the apelike form of the lower jaw led some observers (mostly outside Britain) to question whether the two parts of the skull belonged to the same type of creature. To which suggestion Grafton Elliot Smith responded: "That the jaw and cranial fragments . . . belonged to the same creature has never been any doubt on the part of those who have seriously studied the matter." © *British Museum (Natural History)*

The Piltdown jaw. The forger had filed down the highly cusped molars of the orangutan jaw to give them the flat appearance of human cheek teeth. However, the filing had not been in the same plane, as is obvious from this photograph and should have been obvious to an expert who was examining the specimen critically. © *British Museum (Natural History)*

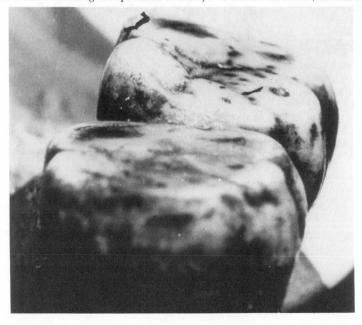

CHAPTER 9

THE KBS TUFF
CONTROVERSY: GENESIS

"This is what caused all the trouble."[1] Frank Fitch is holding up for view a small black-and-white photograph that shows a scatter of unremarkable rectangular crystals measuring less than a centimeter in length. Written on the back is this notation: "*2.42.*" Nothing else.

He taps the face of the photograph ruefully: "These crystals are what led us astray for so long."

The crystals are of feldspar, a potassium-rich mineral common in some volcanic rock. Under the correct experimental conditions, this type of mineral can provide a very accurate indication of the date at which it was spewed, molten and unformed, from the volcano's belly. The 2.42 notation on the back of Fitch's photograph refers to the date at which these particular feldspars are thought to have crystallized from a searing volcanic effluvium in the southern Ethiopian highlands, 200 kilometers or so north of Lake Turkana in Kenya: the date was 2.42 million years ago.

This date is now something of a legend in the annals of paleoanthropology. Mention of it immediately evokes powerful memories in every professional prehistorian, however marginally involved he or she may have been. The episode, known succinctly as "the KBS Tuff controversy," touched virtually everyone, and in so doing cleaved the professional community in two: those who supported the date versus those who thought it was wrong. For more than half a decade—around the mid 1970s—the dispute proved a major distraction for paleoanthropologists, but particularly for Richard Leakey and his colleagues at east Lake Turkana, upon which the controversy was focused.

At one extreme, the KBS affair concerned the esoterica of complex geochronology, which virtually no one understood. At the other extreme was its implication for the antiquity of *Homo*, upon which virtually everyone had an opinion. The middle ground, however, was occupied by the more mundane business of standards of evidence in paleoanthropology. How much do you need to know in order to make a balanced judgment?

The story of the KBS affair will be told in two parts. First, its genesis and substance. Second, in the next chapter, its denouement and impact on the science of paleoanthropology. It is one of those stories in which in retrospect the "right answer" seems perfectly obvious but at the time was obscured by a cloud of uncertainty and vested interest in a particular point of view. It is also a story that demonstrates how very unscientific the process of scientific inquiry sometimes can be.

The seeds of the KBS Tuff controversy were firmly set in the very first full exploratory season at east Lake Turkana, in 1969. That season, remember, was rewarded with the discovery of a complete cranium of a robust australopithecine, *Australopithecus boisei* (KNM-ER 406), the partial cranium (KNM-ER 407) of what at the time was thought to be an early *Homo*, and a cache of stone tools very similar to the most primitive ones found at Olduvai Gorge. Kay Behrensmeyer, a graduate student from Harvard, discovered the tools embedded in the grayish-white ash layer of an ancient volcanic eruption, which immediately offered the prospect of obtaining an accurate date for them from state-of-the-art geophysical techniques. The ash layer thereafter was called the Kay Behrensmeyer Site (KBS) Tuff.

When Leakey saw Behrensmeyer's tools, he immediately recalled having seen similar objects close by during the previous year's excursion to the lake. The terrain in the Koobi Fora area is flat and virtually featureless, ancient sandstones and muds gouged into meadering gullies by ephemeral streams. To the uninitiated it presents a navigational nightmare. But to Leakey, whose spatial memory for such terrain has been honed since childhood, it was a simple matter to recall and relocate the spot where he had seen the tools twelve months earlier. They were about a mile to the south of Behrensmeyer's site. It turned out that the tools that Leakey found were scattered among fossilized hippo bones, and the locality has come to be known as the hippo artifact site. In all likelihood, it was a place where long ago a small band of primitive hominids had butchered the carcass of the lumbering beast, perhaps having chanced upon its dying hulk at the edge of the ancient lake.

Excited by the potential of it all, Leakey immediately contacted Jack Miller, a geophysicist at Cambridge University, England, who specialized in what is known in the trade as geochronology. "A few days ago we located several living sites from which animal remains together with stone tools have been collected," wrote Leakey on 16 June 1969. "Fortunately the tools and bone are embedded in a very promising looking tuff and my geological advisors are exceedingly optimistic about its potential as dating material. If I were to send you a sample of this, would it be possible to get a date run as a matter of some urgency?"

Miller had already been collaborating with Richard's father, Louis, in dating rocks from Olduvai and various other, older sites in Kenya. Richard knew about Miller's work mainly through the recommendation of a mutual friend and colleague, Bill Bishop, a geologist at Bedford College, London. It was therefore natural that Richard should turn to Miller when he needed a volcanic rock dated fast and reliably.

Miller responded immediately to Richard's 16 June letter, saying that he could indeed give the dating work "top priority." Miller worked in collaboration with Frank Fitch, who was a geologist at Birkbeck College, London. The two men had formed a small company, FM Consultants, Ltd., which carried out geochronology for, among other things, the burgeoning North Sea oil-exploration business. They were therefore ready to slot Leakey's sample of volcanic material into their system, for a fairly modest fee.

The exchange of correspondence had been rapid, and by 30 June, Leakey and Behrensmeyer had collected two samples of volcanic tuff from a small hillock a couple of hundred yards to the north of the hippo-butchery site. In his note to Miller, Leakey expressed the hope that the material he was sending would be suitable for the Cambridge dating methods. As matters transpired, it wasn't. The date obtained was more than 200 million years, which quite obviously was way off target. The problem was this. Volcanic ash is ideal for creating an accurate dating framework for a geological sequence, because of the various minerals it contains. And ideally, the ash layers, called tuffs, will form even blankets over the land as they settle to earth after being spewed from volcanic cones. Gradually each deposited tuff is covered with other sedimentary material, and eventually a layer cake of time is created, as volcanic tuffs are interspersed between other rock layers: the oldest is at the bottom, the most recent at the top. But the ideal situation rarely occurs, and particularly so at east Lake Turkana.

Instead of being deposited evenly on the Koobi Fora terrain from the air, the tuffs in the region are formed when the ash from massive volcanic fallout is brought down from the highlands in rivers and streams and spills over onto the surrounding floodplains. The tuffs so formed are often very thick, measuring several meters in depth, but are frequently not pure because they are often churned through older deposits. Contamination with older rocks is therefore an ever-present danger in using material from these so-called reworked tuffs. And so it was with this first sample. The 221-million-year date obtained was that of contaminating "basement" rock.

On receiving this news, Leakey quickly collected and sent two more samples, one of pumice and one of feldspar crystals, the ones featured in Fitch's photograph. This was on 25 July. By 7 August Fitch was able to write back to Leakey, saying that preliminary work on the crystals had given a date of about 2.4 million years, which made him and Miller believe they were dealing with genuine material. Where to proceed from there depended on the choices offered to Leakey, one of which was twice as expensive as the other but, said Fitch, "would result in this tuff being incontrovertibly dated and with greater accuracy than any other site in Africa or elsewhere."[2]

The methods Fitch and Miller were using at the time involved measurements of the elements potassium and argon in the volcanic rock. Potassium contains a very small proportion of a radioactive isotope, potassium-40, which slowly but regularly decays to form the inert gas argon-40. So as time goes by, a rock that contains potassium will accumulate more and more argon-40, thus giving a clock by which the rock can be dated: the more argon-40 there is, the older is the rock. The reason volcanic rock is so appropriate for this kind of dating is that during the eruption all the argon is expelled from the minerals, and the clock is thus reset to zero. Measurement of the argon in volcanic rock therefore tells how long it is since the volcano erupted.

By the time Leakey asked Miller to date this first tuff from Koobi Fora, this so-called conventional potassium/argon technique was well worked out. Miller, meanwhile, had been among a small number of geochronologists who were developing a refinement of the technique, known as the argon-40/argon-39 method. This involves blasting the sample of volcanic material with neutrons, which converts a proportion of an isotope of potassium, potassium-39, to argon-39. Essentially, the measurement of this new argon isotope gives a measurement of the amount of potassium in the sample. And as argon-40 and argon-39 can be determined simultaneously in a machine called a mass spectrometer, it means that the age of the rock can be established with a single experiment, and on very small samples. The conventional potassium/argon technique requires larger samples and the separate measurement of potassium and argon.

The great attraction of the argon-40/argon-39 technique, however, is its enhanced potential sophistication. A simple determination of the age can be obtained by heating of the rock sample to very high temperatures, which releases all the argon at once. If, instead, the heating is done gradually, in steps, then the argon will be released gradually too: the first argon to be released will be that near the surface of the crystal, and as the temperature rises, argon from deeper and deeper within the crystal is expelled. This means that a series of age determinations is produced, essentially giving a profile, or age spectrum, through the crystal. If the crystal has remained entirely unmodified since it first formed, all the age determinations will be the same, and the spectrum will be flat. If, however, the rock has been chemically or physically disturbed in some way over time, so that argon began to leak from the crystal lattice, the first dates obtained will be younger than the ones from the center of the crystal, which may have lost no argon. In this case, the age spectrum will produce an upward slope leading to a plateau. "The technique," says Miller, "is intrinsically more elegant than conventional potassium/argon dating."[3]

Garniss Curtis, a geochronologist at Berkeley who was later to play a major role in the KBS Tuff dating controversy, said in a 1975 publication that the argon-40/argon-39 technique offers greater precision, allows the effects of weathering on a crystal to be "looked through," and permits the investigator to detect any chemical disturbance that might have occurred in a crystal which would be invisible to other methods. He warned, however, that "The interpretation of incremental release diagrams not yielding plateaux is presently very subjective and many differences of opinion have been expressed about them."[4] In other words, unless the age spectrum produced in any particular case was very simple and straightforward, it would not always be possible to understand what exactly it means. This issue, for most of the geochronologists at any rate, was to be at the core of the KBS Tuff controversy.

It so happened, therefore, that Leakey's 16 June request for a dating shot reached the Cambridge laboratory when Miller and his colleagues were justifiably eager to apply the relatively new but potentially powerful argon-40/argon-39 technique to as wide a range of challenges as possible. Leakey simply had to choose between getting a simple, single date using the new technique and the more time-consuming but more sophisticated age-spectrum analysis, the latter of which, according to Fitch, would yield an "incontrovertible" date. Leakey very properly chose this latter course of action.

On 3 September, less than 3 months after the discovery of the artifact-containing tuff, the answer came back: "The age of 2.6 m.y. . . . would appear to be a really good estimate for the age of this tuff horizon," wrote Fitch, a date that was slightly older than the initial 2.4-million-year estimate which had been obtained at the beginning of August.[5] The calculated date was quickly refined to be 2.61 $\pm$ 0.26 million years, which, to anthropologists unfamiliar with the procedures of radiometric dating, has a ring of comforting precision about it. Leakey was of course delighted with the date, especially "in view of the fact that we have artifacts in this tuff."[6] It would be even more significant when, three years later, skull 1470 was discovered *below* the KBS Tuff. Being below the tuff implied that the skull must be considerably older than 2.6 million years, which would make it by far the most ancient member of the genus *Homo* yet discovered. In paleoanthropology in general, and perhaps in the Leakey tradition in particular, such a discovery was extremely important.

Fitch and Miller stuck fast with their 2.61-million-year age (later modified to 2.42 for technical reasons) throughout the controversy. This despite the fact that after the first determination they never again obtained 2.61 from their experiments. For instance, at a conference in Nairobi held in September 1973 they presented 41 separate age determinations on the KBS Tuff, which varied between 223 million and 0.91 million years. Only seven of those 41 measurements came within a quarter of a million years either way of the original 2.61 number, while eight were as close to 1.9. Richard Leakey remained steadfast in his public support of Fitch and Miller's date throughout the affair and abandoned it only in the late 1970s, when its thread of credibility had worn very thin indeed.

The character of the KBS Tuff controversy was in large part colored by the combination of these two factors: Fitch and Miller's solid adherence to their original figure, despite their inability to replicate it adequately; and Leakey's unswerving loyalty to these two men and their contentions. Each party had very good reasons for acting in the way it did. In addition, Leakey clearly had a vested interest in the older date, if for nothing else than because the claim for the oldest *Homo*, oldest stone tools, and so on was good for fund-raising. Always lingering in the background, of course, was the ghost of Kanam. Louis Leakey had suffered a very public ignominy early in his career through his mishandling of dating and geology at that important site, and Richard certainly had no wish to repeat his father's experience.

When Richard Leakey reported in *Nature* the findings of the 1969 expedition, he noted that "The vertebrate material has some similarities with that collected in the Omo Valley by the 1967 International Paleoanthropological Expedition, and detailed comparison between collections should prove interesting."[7] A most prophetic statement this turned out to be, because it was the comparison of certain animal fossils, particularly pigs, between the Koobi Fora and Omo sites that initially drove a wedge between Fitch and Miller's radiometric date of 2.61 million years and the paleoanthropological community's acceptance of it. Very simply, Fitch and Miller's date was out of sync with the story the animal fossils seemed to be telling.

This episode in the dating controversy began with Basil Cooke, a paleontologist at Dalhousie University, Canada, and a longtime associate of Louis Leakey. Cooke was an expert on fossil pigs, which he had studied at Olduvai with Louis and at the Omo. It was natural, therefore, that when the Koobi Fora expedition started to recover some splendid pig fossils, Richard should turn to Cooke for collaboration, which he did in November 1969. In inviting Cooke to work on the Koobi Fora pigs, Leakey was careful to add a condition to the collaboration that reveals his acute sense of the public politics of the science. "My only request on this is that I would like to see the Rudolf area dealt with as a definite locality and not an extension of the Omo project. This could happen although I think it unlikely and I do want to take every precaution as presentation can have a major effect on fund raising work."[8]

Cooke accepted and spent six weeks in Nairobi during 1970, where he saw the many fine specimens that had been collected in the previous two seasons. He also saw that the project had already run into serious problems with its geology. Richard Leakey recounts what happened. "It is essential to relate fossils to the geology of the site where they are found because this provides the vital time reference,"[9] he wrote recently. However, Leakey and his colleagues were so enthusiastic about pursuing the fieldwork that they neglected to ensure that aerial photographs were taken before serious fossil collection began. Without such photographs it is extremely difficult accurately to map fossil finds to the geology of the exploration site. "At the time I was confident about our ability to remember places accurately and I intended that we should mark the spot of each 1970 specimen at whatever time the [aerial] photos became available." This proved to be overoptimistic, and many very good fossils were "lost" in a timeless void. "Some magnificent specimens are of greatly reduced scientific value because of my mistake," Leakey admits.

Nevertheless, there were sufficient pig fossils of known provenance for Cooke eventually to be able to sketch out a picture of the evolution of this animal group at Koobi Fora. But it would take time. Initially he was interested in identifying species and comparing them with animals from other parts of East Africa. During 1970, Fitch and Miller's date for the KBS Tuff wasn't an issue. "There was no particular reason to doubt it,"[10] Cooke recalls.

The problem with the date first began to emerge during 1971, when Cooke was preparing a presentation on the pigs for a conference sponsored by the Wenner-Gren Foundation for Anthropological Research, to be held at Burg-Wartenstein, a magnificent old castle, in Austria. The meeting was titled "Calibration of Hominoid Evolution," so Cooke concentrated on those aspects of the pig fossils which were relevant to time: to wit, the teeth.

Broadly speaking, during evolutionary time the molar teeth of the various species of pig increased in length and height: in effect, the teeth were a paleontological clock. These tooth measurements therefore offered a means of tying in the chronology of separate sites in which pig species occurred, always assuming that evolution proceeded at the same rate in both places. For Cooke, the obvious comparisons with the Koobi Fora pigs implied that the date of 2.61 million years for the KBS Tuff must be wrong, because he judged the pigs found below it to be more like 2 million years old: they matched pigs of this younger age from Olduvai and the Omo. Moreover, there were other animal fossils from beneath the KBS Tuff that appeared nowhere else in Africa earlier than about 2 million years, particularly the modern horse, *Equus*. Cooke was joined in his publication by Vincent Maglio, a young paleontologist from Princeton who was working on various aspects of the Koobi Fora fauna. Here, then, was the first paleontological assault on the radiometric date of 2.61 million years for the KBS Tuff.

At the Wenner-Gren meeting, Clark Howell presented a paper titled "Pliocene/Pleistocene Hominidae in Eastern Africa: Absolute and Relative Age." "That paper contains a tremendous affront to Richard,"[11] Howell now says. "I didn't mean it as an affront at the time, but it clearly was." The paper contained a large chart, with seven columns showing dated units in major fossil-bearing areas. The columns for the area east of Lake Turkana were empty. "The very important successions afforded by exposures in the Ileret and Koobi Fora areas . . . are still under investigation and, as the results are incomplete, their respective columns have been left blank,"[12] noted the legend. Even if it was not an affront, it was certainly a very pointed comment: in effect it said that the dates that Richard Leakey claimed for his fossils were unreliable.

"There was a huge undercurrent at that meeting of there being something wrong with the KBS dating,"[13] remembers Frank Brown, a geologist at the University of Utah, who at the time was a young graduate student. Referring to the impact of Cooke's pig data, Alan Walker recalls that "Basil's numbers were about as shaky as the radiometric dating numbers. His samples were small, and the error bars so large that everything overlapped from top to bottom."[14] In other words, with the relatively limited amount of data Cooke had accumulated by this time, he was unable to be absolutely precise about the figures he was presenting: there was a real element of uncertainty in them, just as there often is with preliminary scientific estimates. This allowed room for subjective interpretation of the data. When the pig data were compared with the apparently solid radiometric date of 2.61 million years,

Cooke's suggestion that it should be more like 2 million failed to make a big impression—in the Leakey camp, that is. A conflict was emerging, but nothing was clear at this stage.

At one point during the meeting, two of the participants leaped onto the tables, grabbed swords that were hanging from the castle walls, and fell upon each other in dramatic combat. "We had practiced the whole thing the night before," says Garniss Curtis, who was one of the swordsmen. "We wanted to liven things up a bit."[15] It wouldn't be long before the KBS dating affair was lively enough without the swordplay.

A year after the 1971 Wenner-Gren meeting, Maglio published on the pigs again, this time in *Nature* and this time in company with data on fossil elephants. But this time he was much more equivocal about what the fossil evidence implied for the date of the KBS Tuff. The paper was accompanied by an anonymous "News and Views" article, "from a correspondent," who applauded the more cautious attitude. The article emphasized the great potential benefit of the more "absolute" methods of dating, and in particular radiometric dating using potassium/argon techniques, such as Fitch and Miller's. "This has been of immense importance in establishing sequences at, for example, Olduvai Gorge,"[16] it opined. It also warned of problems that might arise with the "relative" dating approach of comparing faunas from different geographical locations. "With improved collecting techniques it has become apparent that faunal barriers exist even regionally and can affect distribution of animals in nearby sites. This growing awareness of the obvious errors in attempting correlations based on similar faunal assemblages has latterly made stratigraphers at best dubious of such a dating method."

In the swelling KBS Tuff controversy, which in the first instance at least essentially pitted radiometric dating against faunal correlation, it is not difficult to see which side this correspondent was on: the newly developed potassium/argon dates were to be trusted, while the "old-fashioned" faunal correlations, such as those of Cooke and Maglio, were to be viewed askance.

In addition to throwing his support behind radiometric dating, the author of the "News and Views" commentary also planted the seeds of an idea that later was to find great—but brief—favor among Leakey and his colleagues. The argument was that the animals from the supposed 2.6-million-year time slot at Koobi Fora look younger than those from the same slot at the Omo not because a mistake had been made with the dating of the KBS Tuff at Koobi Fora, but because they were separated by a "faunal barrier" which allowed evolution to proceed at different rates in the two areas. In other words, it was suggested that the animals from below the KBS Tuff at Koobi Fora look more advanced than animals dated at 2.6 million years at the Omo because they were evolving faster. The argument became known simply as "the ecological hypothesis," and was propounded vigorously by Leakey, Behrensmeyer, and their colleague John Harris, who was a paleontologist on the Koobi Fora project.

"Yes, it sounds pretty stupid when you think properly about it,"[17] Harris now says. "Our position at the time was that Fitch and Miller's number for the KBS Tuff is a good date, well established by geochronology. Our science—paleontology—is interpretive, so we have to look for other explanations of the apparent discrepancy between the faunas. I was therefore amenable to the faunal-barrier idea. I can now see that we were seeking ways of justifying the date rather than objectively trying to clarify the evidence."

There are differences in the modern ecological communities at Koobi Fora and the Omo, which is perhaps not surprising, as the two are separated by the huge lower Omo River. But whether this would be sufficient to permit separate rates of evolution is problematical. Ernst Mayr, one of the great modern evolutionary biologists, is quite clear about the issue. "One can have rather different rates of evolution on islands, particularly if the populations there were established as founder populations,"[18] he says. "But rates of evolution of such difference on continents are most unlikely, indeed I would say, are quite unheard of."

One reason the ecological hypothesis flourished among Leakey and his colleagues in Kenya was their separation from modern scholars in evolutionary biology. "We were pretty isolated in Nairobi,"[19] says Harris. "Most of the people I saw were part of the Koobi Fora team, who subscribed to the same sorts of ideas. We were convincing ourselves that we were right."

Leakey's position at this point—during 1972 and 1973—is clearly reflected in his April 1973 *Nature* paper in which he announced the discovery of the famous 1470 skull. The skull, he explained, had been recovered from below the KBS Tuff, "which has been *securely dated* at 2.6 m.y." [Emphasis added]. No sign of equivocation here. He believed that the fossil collection at Koobi

Fora was not sufficiently developed for clear comparisons with the Omo fauna. He also frequently said that the radiometric-dating framework at the Omo might itself be wrong, which would invalidate correlations with Koobi Fora. And, like most paleoanthropologists unfamiliar with geochronology, he could see no reason to doubt Fitch and Miller's argon-40/argon-39 date. "Frank and Jack are very convincing people,"[20] he now says. "If you have no knowledge of geochronology, as I certainly did not, and you see that these people come from one of the best labs in the world, you think they must know what they are doing."

Toward the end of 1973, however, the pressure was beginning to build, and it exploded right in Leakey's backyard. At yet another Wenner-Gren conference, held in Nairobi 9 to 19 September 1973, the rivalries that had been simmering steadily over the previous couple of years came clearly to a boil. There developed a stark polarization between those who supported the 2.61 date and those who were against it; those who worked at Koobi Fora and those who worked at the Omo; those in Leakey's team, who were loosely tied to Nairobi, and those who were allied with Clark Howell, who was based in Berkeley. Howell, who had questioned the validity of Fitch and Miller's radiometric date for the KBS Tuff at the 1971 Wenner-Gren meeting, was one of the scientific leaders of the Omo expedition. In the eyes of Leakey and his followers, Howell was therefore the boss of the rival gang. Leakey had been part of Howell's Omo expedition during 1967, but had left when he discovered the potential of Koobi Fora. The sense of rivalry between the two groups can only have been sharpened by this aspect of the history.

Shortly before the 1973 Wenner-Gren meeting Leakey gathered his team for a party at his house in Karen, a suburb of Nairobi. "There were a lot of us there,"[21] recalls Michael Day. "There was a tremendous sense of 'We are here to defeat the other side.' "

Clark Howell, meanwhile, was becoming more and more convinced that Richard Leakey had a serious problem with the KBS Tuff dating. Just before the Nairobi meeting, he was going over the issues with Frank Brown while they were still in camp at the Omo. "You can't have all these things that just don't fit between the two areas that are just across the river from each other,"[22] Howell said to Brown. "One set or the other of these numbers is wrong. How sure are you of your numbers for the Omo, Frank?" "Well, as sure as I can be," Brown replied. "I don't think there's anything wrong with them." "Well," Howell said finally after a long pause, "then there's got to be something wrong with the KBS date."

As a result of this conversation, Howell gave an extraordinary paper at the Nairobi meeting. He didn't indulge in long arguments about the problems with the dating. He didn't try to suggest what was wrong. He simply read out long lists of fossil species, one set from below the KBS Tuff at Koobi Fora, which was supposed to be at least 2.6 million years old, and the other from strata of equivalent age from the Omo. He simply said, "They are not congruent," and left it to his audience to draw their own conclusions. If the geological sections Howell was comparing from Koobi Fora had been the same age as those from the Omo, the two lists of fossil animal species would have been very similar. That they were not was clear from Howell's presentation, and the conclusion—that the dates were different, with the Koobi Fora section being younger than Leakey and his colleagues contended it was—should have been blindingly obvious to the objective eye. But since most people had gone to the meeting prepared to protect their conclusions, not change them, Howell's paleontological tour de force had little impact.

While Leakey was not prepared to be swayed by Howell's presentation, he was quietly impressed with the arguments of Alan Gentry, a paleontologist from the British Museum (Natural History) in London. Gentry is a gentle-mannered man, not given to jumping into controversies with unfounded assertions. When he said he thought that Fitch and Miller's date for the KBS Tuff was too old because beneath the Tuff was evidence of a certain antelope, which is dated at about 2 million years at Olduvai, people listened. And this included Leakey and particularly his co-leader, the late Glynn Isaac.

Isaac was in something of a difficult position throughout the KBS affair. An archeologist of worldwide distinction, Isaac helped Leakey run the Koobi Fora research program, and was therefore firmly identified with the Fitch/Miller position—and yet he was a professor at Berkeley, site of the enemy camp. Eventually, his intimacy with both camps would help untangle the dating problem. Although Gentry's evidence planted seeds of doubt in Isaac's mind, he was vigorous in his defense of the older date throughout the meeting. This led him at one point to make his now-famous remark that what the Koobi Fora people needed was "pigproof helmets"—referring to Cooke's ever-more-pertinent fossil analysis. Isaac's remark was meant to lighten an atmosphere that had grown distinctly tense in the sharpening rivalry of the conference, but it clearly revealed his campaign stance. It also irritated Cooke, Howell, and their colleagues, who considered that their efforts were being trivialized.

Cooke had once again delivered the message of the pigs, which was essentially the same one he had presented two years earlier, but bolstered by more data. From the information he now had, he said, "it would be inferred that the KBS Tuff should be fairly close in age to the top of Member F at Omo, which is apparently 2.0 m.y., whereas the radiometric date for the KBS Tuff is 2.6 m.y. The discrepancy is considerable and cannot be ignored."[23] Cooke's manner was always calm, but his message was delivered with unambiguous clarity. It was a message that did not change in essence from beginning to end.

Postconference commentary highlighted the complex of geophysical and paleontological problems faced by researchers. Bill Bishop, for instance, noted that at Omo there were about 120 identifiable tuffs, whereas the Koobi Fora people counted only 15. Something must be wrong here, he thought. Referring to Cooke's "considerable discrepancy," Bishop indicated, however, that it should not necessarily be cause for alarm. "I believe that it would be remarkable and perhaps even 'suspicious' if such imprecise methods of correlation as those depending upon stage of evolution of mammalian taxa or even statistical analysis of mammalian assemblages yielded results identical with those based on isotope chronometry."[24] Karl Butzer, another scholar on the sidelines, noted accurately of the meeting that "Interpretation of the geology proved to be most controversial."[25] Like many paleoanthropologists, however, he was impressed with the radiometric dating. "These East Rudolf argon-40/argon-39 determinations are age spectra datings, the best of their kind," he said.

Of all the observers of the KBS controversy, Bishop was one of the more objective, and Leakey and his colleagues could therefore take justifiable comfort in his words. And Butzer's comments naturally served to bolster Leakey's position even further. "Yes, even though it had been a lively meeting, with the Omo people disagreeing with us a good deal, I felt we had good reason to be pretty secure with our date."[26] remembers Leakey.

In fact, by the time of the 1973 Nairobi meeting, Fitch and Miller had become quite troubled with the results they had been getting with the dating material from Koobi Fora. They had carried out determinations on more than half a dozen volcanic tuffs, many of which produced satisfactory dates. But the KBS Tuff was particularly errant: it gave a startling scatter of dates, as mentioned earlier. Accounting for dates older than 2.6 was easy: contamination by older volcanics. But what of dates younger than 2.6? How could this be explained?

Miller is very firm about the reliability of his experimental techniques. "There is no such thing as a bad date,"[27] he asserts. "The numbers you get are trying to tell you something—assuming you've got the thing right." He and Fitch felt confident about the 2.61-million-year date from the original crystals, because they were so obviously genuine. "So we had to invent a model to account for the scatter of results," recalls Miller. "I talked with some of my chums around here, and one of them pointed out to me the rather obvious thing that in the Koobi Fora area, with its warm alkaline environment, these volcanic minerals are likely to undergo some fairly fundamental chemical changes." And so was born the "overprinting" explanation for the younger dates.

Overprinting, which is a fairly uncommon phenomenon in geology, refers to specific events of heat, pressure, or chemical environment (or any combination of them) that alter the composition of the exposed minerals in some manner. For instance, feldspars that are altered sufficiently will begin to lose their argon, and will therefore register as being younger in potassium/argon experiments. "It was a good working hypothesis for the KBS Tuff,"[28] Miller now says. "It seemed a sensible mechanism." In other words, if the KBS Tuff had been exposed to some hydrothermal event, say 0.7 million years after it was deposited, it could give an age determination of 1.9 million instead of 2.6. In effect, overprinting resets the radiometric clock and leads to an erroneously younger date. Fitch and Miller developed and promulgated the idea that there had been a whole series of overprinting events, including one at 1.9 million years ago, which could account for all the figures they obtained that were less than 2.6 million. They included this explanation in their paper for the Nairobi conference.

Just as papers to scientific journals are subject to review by experts in the field, so too are many of those destined to be published in "conference volumes." Fitch and Miller's paper initially went to two reviewers, Brent Dalrymple of the U.S. Geological Survey, Menlo Park, California, and a geophysicist at the Smithsonian

Institution's Astrophysical Laboratory in Cambridge, Massachusetts. The two opinions were, to put it mildly, contradictory. Dalrymple said that the paper "does not meet the normal scientific standards of evaluation of data."[29] His specific complaints were that the paper did not present experimental data, just conclusions drawn from them; and the plotting of age spectra was not properly quantitative, so that other scientists would not be able to attempt to replicate the work. Being an expert in the evaluation of argon-40/argon-39 age spectra, Dalrymple stated that "it is my opinion that the interpretations of age spectra in the Fitch-Miller paper are untested hypotheses, not proven facts."

Dalrymple was also unhappy about Fitch and Miller's invoking both contamination and overprinting to account for dates other than 2.61. "These two mechanisms could be used to explain anything, as their effects on the potassium-argon technique are exactly opposite." Referring then to the context in which the paper was being used, Dalrymple considered that "In its present form it could seriously mislead scientists who have a critical interest in the subject but are not specialists in geochronology." Richard Leakey, for instance.

Reviews of scientific papers are often anonymous, but following his usual practice, Dalrymple said, "I insist that my identity be made known to Frank and Jack."

The second review took an entirely different stance. "Fitch and Miller," it said, "are pushing the argon-40/argon-39 technique ahead in new fields, while at the same time explaining carefully what they are doing, and being reasonably prudent where new judgments have to be made."[30]

Faced with conflicting opinions, the editors of the conference sought a third opinion. They approached Ian McDougall of the Australian National University, Canberra, who is one of the most respected potassium/argon geochronologists in the world. "There is no doubt that it is extremely difficult to obtain reliable potassium-argon ages on these rocks, and I do not wish to deprecate the attempts made by Fitch and Miller,"[31] he wrote. He then went on to express disappointment in the paper and enumerated precisely the same shortcomings identified by Dalrymple. "In summary, in my view this paper does not even approach the standards required of a scientific journal."

But the Fitch-Miller paper was subsequently published without significant alteration from the version seen by Dalrymple and

McDougall. The editors of the conference volume were Yves Coppens, Clark Howell, Glynn Isaac, and Richard Leakey. Isaac was designated as editor of the section within which the Fitch and Miller paper was contained. Exercising his option to take note of, but not necessarily act on, reviewers' opinions, Isaac elected to accept Fitch and Miller's paper more or less as submitted. To do otherwise would have threatened a mighty upset in the Leakey camp.

Fitch and Miller exchanged opinions with Dalrymple following the latter's rather negative review. Their principal defense was that they had accumulated a "wealth of experience . . . in the hard school of commercial geology."[32] Dalrymple was not impressed. "Neither I nor any other scientist is obliged to accept conclusions based on privileged information not freely available in the published literature," he responded. "It is not incumbent upon me to prove that you are wrong, it is your burden to prove that you are right. You are entitled to offer your hypotheses but beyond that you are required by the scientific method to present conclusions only when you can offer reasonable proof that they are correct."

Fitch and Miller now say that they were rather surprised by Dalrymple's negative tone. Nevertheless, says Miller, "Dalrymple has always held that only perfect and unaltered rocks can be reliably dated. Most rocks have suffered alteration to some degree and to take such a dogmatic view would both limit the application of the method and sidestep the intellectual challenge that such material presents."[33] Miller goes on to argue that experience has shown that "imperfect" rocks can be reliably dated with the age-spectrum technique, and that at the time in question, "we had something like 10 years of experience of argon-40/argon-39 dating."

On the question of absence of data in their paper, Miller explains that "this was a restriction put on us by the editors." Perhaps Fitch and Miller were unlucky with their editors, because according to Dalrymple, "During the period in question, Jack and Frank were the only people I can recall leaving out data from their papers and using nonquantitative plots. I never did figure out why they did this and never got a sensible explanation from them either."[34] The absence of data and the nonquantitative way age spectra were plotted meant that "no one could check their calculations, reproduce their measurements, or interpret their results."

Argon-40/argon-39 data are often voluminous and can consume

precious space in journals and conference volumes. "The great bulk of analytical data produced by argon-40/argon-39 dating had already made many editors refuse to accept it in full,"[35] explains Fitch. "Our usual response to this problem was to offer copies of the full data to interested parties on request. This we did in Nairobi."[36] The invitation was included in the published version of the paper. Dalrymple's reaction to this is that scientists should not have to resort to such tactics in order to have access to essential information. "A scientific publication is, after all, a permanent archive of an experiment, an investigation, a hypothesis, or a theory."[37]

McDougall recalls a visit to Cambridge, England, in September 1977: "I was made most welcome, and Jack kindly had me stay at his home. . . . [However], *No* Kenyan data were shown to me, despite repeated requests to see examples of their primary data from East Africa or other projects, and I had great difficulty in obtaining much information concerning their techniques. They were unduly defensive and I came away with a feeling of concern as to what was going on."[38]

Miller's response to all this is typically resolute. He recently said, "The age spectra clearly demonstrated overprinting and we had firm evidence to demonstrate contamination in some samples."[39] Period. This line of argument was solidly maintained from the time of its inception until the whole affair atrophied in the early 1980s.

The September 1973 Nairobi conference was followed by a similar gathering in New York in January 1974. Leakey was proud to show slides of the splendid hominid fossil skulls and jaws that had been collected from Koobi Fora, which made Clark Howell's fragments of jaws and teeth from the Omo look pretty meager by comparison. Observers interpreted Leakey's actions as his way of exacting some revenge on Howell for what had transpired in Nairobi. Cooke recounted once again the story of the pigs, which, once again, incensed Leakey, so much so that Leakey later felt constrained to write an apology to Cooke for being "relatively harsh." Cooke did not feel particularly affronted, merely that Leakey "was being unduly obstinate in the face of the evidence."[40]

The year 1974 saw the publication of several papers that independently appeared to support the Fitch-Miller postion. One was a description of results from a different method of geological dating—paleomagnetic reversal—by Glynn Isaac and Andrew Brock.

Based on the fact that the earth's "inner magnet" occasionally reverses, making the north magnetic pole into the south and vice versa, the technique can provide something of a clock for rocks that can become magnetized. According to this clock, said Isaac and Brock, a date of about 2.6 million years seemed reasonable for the KBS Tuff. A second technique—fission-track dating, of which more later—also bolstered the chronology favored by the Koobi Fora camp. Indeed, of the dozen or so papers published by this time in the scientific literature that directly related to the age of the KBS Tuff, only two explicitly said that the 2.6-million-year date was wrong. And both of these were by Basil Cooke, both based on the same fossil pig evidence. So although a strong undercurrent had developed that "something was wrong" with the 2.6-million-year date for the KBS Tuff, the scientific literature overwhelmingly supported it.

But 1974 was a turning point for the Fitch-Miller chronology, because a second potassium/argon laboratory began to insinuate itself into the act.

And it was an ironic event. Garniss Curtis had worked with Louis Leakey some years earlier, and had done the original dating at Olduvai Gorge in 1960. Not long afterward, the two men had a mighty row over the dating of some important sediments in western Kenya, where Louis had found fossils of what he thought might be man's earliest forebears. Curtis' radiometric dating put the rocks at about 17 million years, while Louis contended, from the evidence of the other fossils at the site, that they were twice that old. "Louis wanted those rocks to be old, because of his belief in early *Homo*, but I knew they were much younger,"[41] remembers Curtis. The younger date turned out to be right, but by the time this was proved Curtis and Leakey had parted company because of their disagreement, and Curtis vowed never again to set foot on the African continent while Louis Leakey was alive.

Leakey needed a geochronologist to replace Curtis, and eventually invited Jack Miller to join him. He accepted, and the two men collaborated for several years. It was natural, therefore, that when in 1969 Richard Leakey needed a geochronologist for his Koobi Fora site he should turn to Miller. Had Curtis and Louis Leakey not split up after their confrontation, Curtis would still have been working in East Africa, and Richard Leakey might well have asked him to do the initial dating at Koobi Fora. As it was, Curtis had caused a storm in the older Leakey generation by producing an unacceptably young date, and he was about to do the same thing with the next Leakey generation.

The precise means by which Curtis came to do his conventional potassium/argon dates on the KBS Tuff material have always been the subject of conjecture and scurrilous rumor. There never was a formal invitation from Richard Leakey, for instance. And the manner in which the results eventually came out only served to fuel rumors of intrigue.

What in fact happened was that Glynn Isaac, becoming more and more uneasy about the Fitch-Miller date, thought it would be valuable to seek a second opinion. In a series of casual conversations with Curtis, whose laboratory was on the opposite side of the Berkeley campus from his own, Isaac suggested Curtis might like to have a go at the infamous tuff. Thure Cerling, a student of one of Curtis' colleagues, was already involved in exploratory geochemical investigations at Koobi Fora, so it would be easy for him to make collections of volcanic material from the KBS Tuff that would be suitable for potassium/argon dating.

Cerling left for Kenya in June 1974, and called on Fitch and Miller in England on the way. "I spoke with Frank Fitch concerning the possibility of Garniss dating samples from Koobi Fora that I would collect that summer,"[42] Cerling recalls. "I suspect he had not been approached by Glynn, for he did not indicate any previous knowledge of the Berkeley group getting involved. He indicated that it was OK with him."[43] Miller reacted likewise. "I would certainly not expect Garniss to have to ask my permission to date the tuff,"[44] Miller now says. "Science is free, and it should be open to anyone who has any interest in doing it." Fitch now comments: "I certainly didn't object. The more knowledge one obtains, the better."[45]

Once he reached Koobi Fora, Cerling began making collections of KBS Tuff samples from several locations. "There were a lot of people around who knew what I was doing."[46] he recalls. Frank Fitch came to the field in August, and was most impressed with the quality of Cerling's samples. "I hadn't seen anything like it before,"[47] he now says. "It was a better sample than we'd ever had."

Later on there would be numerous rumors around Nairobi that Cerling had made the tuff collections secretly, that he'd hidden the rocks under his bed, and had successfully foiled attempts by unnamed Leakey-camp members to steal them. "No, I had no problems at all with the collection,"[48] Cerling now says. "I don't know how those rumors started, but they weren't true at all." The existence of these rumors, though unfounded, is a cogent indication of the tense atmosphere that prevailed in Kenya at the time. "The only slight hitch was when I got back to the main Koobi Fora camp after Frank had gone back to Nairobi. He had left a handwritten note with Glynn, who gave it to me. The note simply said that he would prefer to have a written request from Garniss to go ahead with the date. Just courtesy, I thought. I said, 'OK, I'll tell Garniss.' Garniss said, 'That's fine with me; I'll write a letter'."

At this point the story becomes murky. Curtis claims he wrote to Fitch, who denies ever receiving such a letter. Perhaps Curtis put it in the wrong envelope or something, Fitch suggested. "Anyway, there was no initial response to this letter of mine, so I said we should do the potassium estimations but wait to hear from Frank before doing the argon determinations,"[49] remembers Curtis. "This way we could push on with the work without actually getting a date without their permission." The problem was eventually resolved; but by this time it was January 1975, just weeks before a major meeting that had been planned by the Geological Society of London. It was to be called "Geological Background to Fossil Man," and was to focus on East Africa. The KBS Tuff affair would inevitably be prominent in the proceedings. Leakey would be there. Fitch and Miller would be there. And so would Clark Howell. Curtis wanted to be there too, and he wanted to be able to present his date for the KBS Tuff.

With just days to spare, Curtis and his colleagues completed the argon analysis, computed the calculations, and got their date—or rather, their dates. It seemed that not only had Fitch and Miller been clinging to too old a date for the KBS Tuff, but also there were two tuffs in what was called the KBS, not one. One of them, in Curtis' analysis, was 1.6 million years old and the other 1.8. Armed with these results, Curtis left for England and on Wednesday, 19 February, arrived at Burlington House, in London's Piccadilly, which contains the grand meeting rooms of the Geological Society. Curtis had not thought it necessary to contact Leakey with his results. Nor had he had time to discuss them with Fitch and Miller. Nevertheless, news of Curtis' results had quickly spread along the grapevine in the few days prior to the meeting— a sequence of events that erroneously added further to the air of secrecy in which it was thought that Curtis had been working.

Meanwhile, toward the end of 1974 Leakey contacted Miller to express his concern about the upcoming Geological Society meeting and to urge a firm approach. "There is every indication that Berkeley is sending a 'team' to do us in at London on the issue of dating,"[50] he wrote. "I am more than prepared to settle the issue regarding fauna and you will have a chuckler. I am sure you will deal with the geophysical issues and I merely urge that we keep it cool and 100% effective." Miller concurred. "I am sure that we shall have some good stuff for February and I am in one hundred per-cent agreement with you that we should be completely cool about the job; I have always thought that the thing was getting too hysterical."[51]

Just a few days before the February meeting itself, most of the members of the Koobi Fora research group gathered at Miller's laboratory in Cambridge, where they discussed the events of the previous year's research and the upcoming meeting. Naturally, news of Curtis' 1.8-million-year result for the KBS Tuff was a hot topic of conversation. "Frank told us he didn't believe it," remembers John Harris. "He said that the 2.6-million-year date was secure."[52] The group agreed that its best plan of action against the anticipated attack from Berkeley was to play it cool, just as Leakey had earlier urged Miller at the end of November.

Ales Hrdlicka, a Czech immigrant to the United States, became a powerful figure in American anthropology in the 1920s and '30s. He effectively destroyed G. Edward Lewis' claims for *Ramapithecus*, saying that Lewis' paper contained "a series of errors" and had reached an "utterly unjustifiable" conclusion. © *Smithsonian Museum of Natural History, Washington, D.C.*

G. Edward Lewis, seen here in the Siwalik Hills, India, in 1932, the year he found the first *Ramapithecus* specimens. Of Hrdlicka, he now says that "he thought he was the anointed and elect prophet who had been foreordained and chosen to make such discoveries and demolish the work of anyone else." © *G. Edward Lewis*

Ramapithecus: type specimen (two pieces of lower jaw, near center of picture) displayed on a field map of the Siwalik Hills and G. Edward Lewis' field notes. © *John Reader*

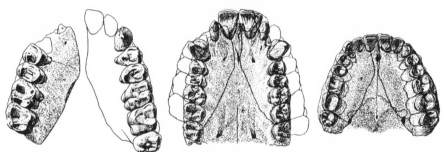

Reconstructed jaw of *Ramapithecus*. This diagram appeared in a 1964 *Scientific American* article by Elwyn Simons. It shows, left to right, fragments of the upper jaw of *Ramapithecus* as reconstructed by Simons; an orangutan's upper jaw, with the outline of the *Ramapithecus* fragments superimposed; and a human upper jaw, with the same outline superimposed. The caption read: "The U-shaped arc formed by the ape's teeth contrasts sharply with the curved arc in *Ramapithecus*, which is closer to the human curve." A complete upper jaw of *Ramapithecus* discovered later, however, showed the shape to be more a V than an arc. © *Scientific American, July 1964.*

Vincent Sarich (right) with Sherwood Washburn, who encouraged Sarich to begin working on the molecular clock for human origins. As a result, Sarich once said, "One no longer has the option of considering a fossil specimen older than about eight million years a hominid *no matter what it looks like.*" Although he agreed with Sarich's conclusion, Washburn considered that "It was the dumbest thing he ever said." Christopher Springman
© Discover *magazine*

Elwyn Simons (right) and David Pilbeam in Rome during May 1982, attending a Pontifical Academy meeting on human origins. Later that year the two men were studying the new fossil evidence from Pakistan. "Eventually I said to David, 'This is a convincing tie between *Sivapithecus* and the orangutan.' We both knew what that meant," says Simons.
© *D. Pilbeam*

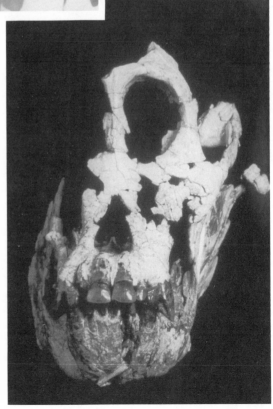

The *Sivapithecus* face from Pakistan: the fossil evidence which confirmed that what the molecular biologists had been saying was indeed correct.
© *W. Sacco*

Allan Wilson, Sarich's colleague in the new thrust on the molecular clock, was dismayed by Morris Goodman's suggestion that the clock had slowed down in primates. "I'm pretty intolerant of the idea of a slowdown," says Wilson. "Why does he say it? He says it is because he is afraid to take on the paleoanthropologists."
© *University of California, Berkeley*

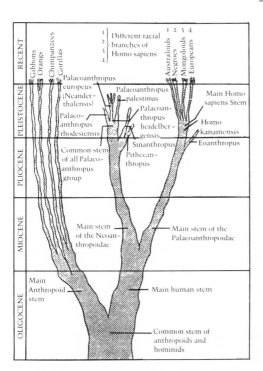

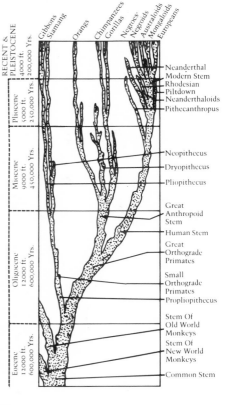

Human family trees, by Louis Leakey (left) and Sir Arthur Keith—the former's in *Adam's Ancestors*, published in 1934, and the latter's in *The Antiquity of Man*, first published in 1915. Leakey was greatly influenced by Keith, and this is reflected in both the style and the content of their depictions of human evolution. Both men showed the human line splitting from the apes in the Oligocene; both men put Piltdown (*Eoanthropus*), Neanderthal, and *Pithecanthropus* on side branches; and both men indicated long histories for the modern races of mankind.

Louis Leakey (center) with Mary Leakey and Peter Kent at Olduvai Gorge in 1935. "I was born in East Africa, and I've already found traces of early man there," Leakey told a Cambridge don who was puzzled by Leakey's interest in Olduvai. "I'm convinced that Africa, not Asia, is the cradle of mankind," he said, which ran counter to the professional sentiment of the time. © *Leakey archives*

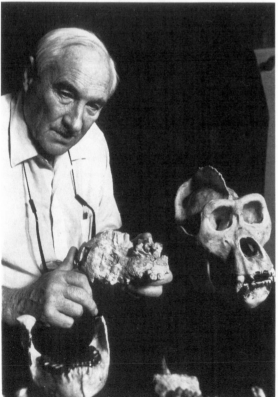

Louis Leakey holds one of the *Homo habilis* fossils from Olduvai Gorge. A Zinj-like jaw is in the foreground and a gorilla skull to the right. Leakey quickly formed the impression that the new fossils were different from Zinj. "Mary and I are sure . . . that it is NOT *Australopithecus*," he wrote to his colleague Phillip Tobias in December 1962. "Only those with 'Psychosclerosis' . . . could ever put it in that sub-family." © *L.S.B. Leakey Foundation*

Olduvai Gorge, on the southern edge of the Serengeti Plain in Tanzania, where Louis and Mary Leakey spent more than three decades in the search for ancient *Homo*.
© *University of California Press.*

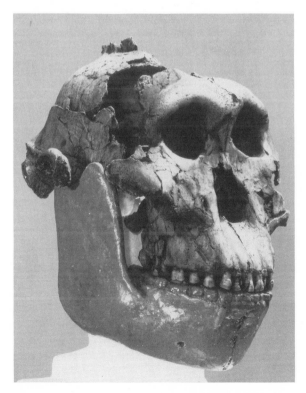

Zinjanthropus boisei, the skull that was found by Mary Leakey in July 1959 and propelled the Leakey name to worldwide fame.
© *Bob Campbell*

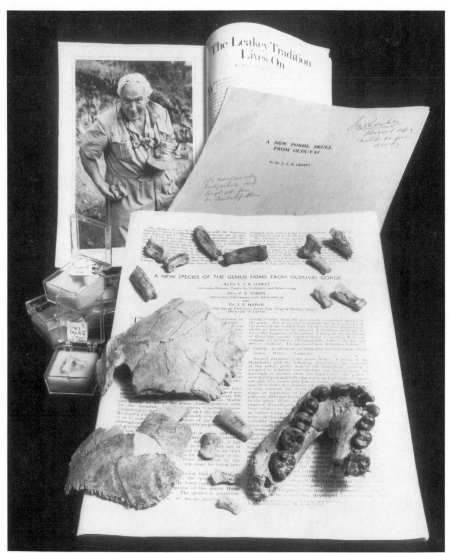

Homo habilis: the type specimen, displayed on a copy of the April 1964 *Nature* paper in which the new species was announced. A critic of Leakey's, Sir Wilfred Le Gros Clark, said in response to the paper: "One is led to hope that ["Homo habilis"] will disappear as rapidly as he came. . . . It certainly does not appear to merit prolonged controversy." The controversy in fact continues to this day. © *John Reader*

The Leakey family in the 1950s. From left to right, Richard, Mary, Philip, Louis, and Jonathan, in company with the family's Dalmations. The Leakey children were always being dragged off to remote spots in search of fossils, which made Richard resolve at a young age not to have anything to do with paleontology when he grew up. The resolve was not to last. © *Leakey archives*

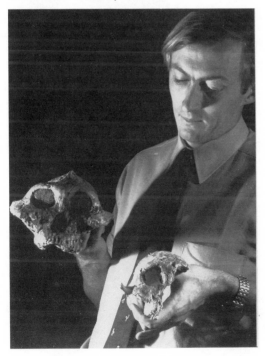

Richard Leakey holds in his right hand the robust australopithecine cranium (number 406) that he found in 1969, the first full season at Koobi Fora. That discovery dramatically changed Leakey's outlook. "With 406, the emotional charge I got from finding something myself, something everyone was going to like, was extremely powerful," recalls Leakey. "I suppose it triggered off this paleontological possessiveness we all experience." © *Bob Campbell*

Louis and Richard Leakey discuss a fossil during the 1967 international expedition to the Omo Valley, Ethiopia. Richard Leakey now says of himself at the time: "I had no preconceptions about human evolution, nor any desire to postulate a particular idea in terms of family trees. I was very much under Louis' guidance. I didn't presume to have an opinion." © *Bob Campbell*

Meave and Richard Leakey, holding the famous 2-million-year-old 1470 cranium and a hominid thighbone found near the same site in August 1972. "I feel confident that one day we will be able to follow man's fossil trail at East Rudolf back as far as four million years," said Richard Leakey shortly after the discovery. "There, perhaps, we will find evidence of a common ancestor for *Australopithecus*—near-man—and the genus *Homo*." © *Bob Campbell*

Excavation at Koobi Fora of a *Homo erectus* cranium in 1977. Left to right: Glynn Isaac, Jack Harris, Richard Leakey, Meave Leakey, Kamoya Kimeu. Not long afterward, Leakey's personal and professional lives were threatened by medical and other problems, and he contemplated giving up fossil hunting for good. Leakey's colleagues persuaded him against such a decision, "So I stayed with it, and I'm enjoying myself again," he now says.
© *R. Lewin*

Donald Johanson (left) and Richard Leakey at a scientific meeting in Philadelphia, February 1979, shortly after the announcement of Lucy. "Richard Leakey, the Kenyan anthropologist, is challenging the announcement made last month by two American scientists that they had discovered a new species," ran a report in *The New York Times*. "Although honest differences of opinion are common enough in all sciences, there are overtones of a confrontation between the two anthropologists." *John Alexandrowitz* © *NYT Pictures*

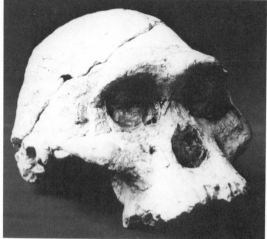

Parade of hominid fossils:
A. *Australopithecus africanus*
("Mrs. Ples") from Sterkfontein,
South Africa.

B. *Australopithecus robustus*
from Swartkrans, South Africa.

C. "Partial skeleton" of
Australopithecus africanus,
from Sterkfontein.

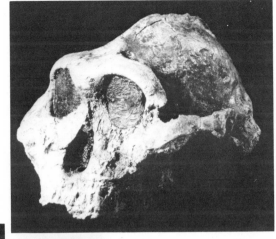

D. *Australopithecus boisei* from Koobi Fora, Kenya. This is cranium number 406, the first one found at Koobi Fora.

E. *Homo habilis* from Koobi Fora. This is cranium number 1470.

F. *Homo erectus* form Koobi Fora. *P. Kain   © Sherma*

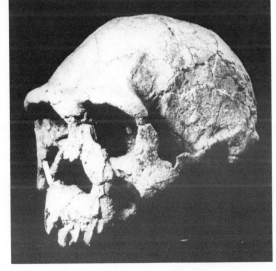

Fossil hunting at Koobi Fora, where "the hominid gang" prospect for fossils by systematically walking many hundreds of square kilometers of terrain each year. © R. Lewin

CHAPTER *10*

THE KBS TUFF CONTROVERSY: DENOUEMENT

E ven though fewer than a quarter of the papers delivered at the February 1975 Geological Society meeting were directly concerned with the age and implications of the KBS Tuff, the issue completely overwhelmed the atmosphere of the occasion. For four years the controversy had steadily fermented, and an initial unease eventually turned into outright and public confrontation. The Nairobi conference of September 1973 had represented the first real eruption of the conflict. Now, in the elegant marbled halls of the Geological Society meeting rooms, the issue was about to be blown wide open.

Garniss Curtis was due to speak at the end of the second day of the conference, on Thursday the 20th of February, by which time his message was well known to his audience. There was much gossip around the meeting about how, supposedly, Curtis had clandestinely obtained his samples and had performed the experimental work in complete secrecy. There was talk, too, that the rock he had dated was not from the KBS Tuff at all, but had been mistakenly collected from a younger tuff, hence his "erroneous" date of 1.8 million years. Most of all, members of the Leakey camp simply dismissed Curtis' work as just plain wrong, and they tried to brush it aside as if it were of little account.

However, the degree to which the new attack from Berkeley

really wounded Leakey and his colleagues became apparent on the evening following Curtis' paper, when Leakey invited a number of people to dinner where he was staying, in Hyde Park Square. "We had a fight," recalls Howell. "I never saw anything like it. Richard and I tried to cool it off."[1] The people there were something of a mixture; they included, in addition to Howell, Bill Bishop, Glynn Isaac, Bernard Wood, Michael Day, Frank Brown, Kay Behrensmeyer, and Don Johanson. Nearly everyone was involved in the KBS Tuff controversy, and all had strong opinions.

Discussion, all shop talk, ranged over several topics, but inevitably it kept coming back to the geology and fauna of Koobi Fora, with the KBS Tuff at its focus. "In the end it got quite heated, and proper argument eventually gave way to 'You are completely wrong,' to which the reply was 'No, you are completely wrong.' Sounds like faith, not science, doesn't it?" says Howell.

At one point, Leakey produced a fiberglass cast of part of a hominid pelvis that had been discovered the previous season. "Just like Louis used to do, with a flourish," recounts Howell. Leakey argued that the pelvis, which looked very modern and was certainly *Homo*, had been recovered from below the Tulu Bor Tuff, which would make it more than 3 million years old. So, he said, even if 1470 turned out not to be as old as 2.6 million years, if the KBS Tuff was wrongly dated, he still had early *Homo* from Koobi Fora. Here, then, was a glimpse of the issue underlying the KBS Tuff controversy: namely, proof of ancient *Homo*.

Howell was skeptical of Leakey's interpretation of the cast of the fossil pelvis; he said, "That's just like Olduvai hominid 28, from Bed IV.' " He was referring to a *Homo erectus* pelvis from Olduvai Gorge, dated at about a million years old, thus throwing doubt on the validity of Leakey's specimen. "Michael Day looked at it, and hemmed and hawed, saying he wasn't sure. I said, 'Come on, Michael, you know morphology, that's your job. You are the one who described OH 28. What are you talking about?' That started another fight." Leakey joined in and said, "It proves we have *Homo* at 3 million years. Yes, it's similar to OH 28, but I assure you it's different. I've looked at it." Howell replied, "Well, look again. I just don't believe your 3-million-year-old *Homo*. And I don't believe your 2.6-million KBS Tuff." Everyone got into that fight. "It was a very stormy occasion," says Howell.

It turned out later that because of endemic confusion in the Koobi Fora geology, the pelvis had in fact come from much higher up in the succession than was initially thought, and was therefore much younger: more 1.9 million rather than 3 million years. However, Howell had been right and Leakey wrong: the pelvis was indeed like the *Homo erectus* specimen from Olduvai Gorge and was not a more primitive, more ancient version of *Homo*. By this time, however, Leakey and Isaac had already reported once in the scientific literature that the partial pelvis provided "dramatic confirmation"[2] of ancient *Homo* at Koobi Fora. This appeared in a little-circulated volume of papers published in honor of Louis Leakey. The error in the geology was discovered just as a second paper was about to be published, this time in *Nature*. It was hastily withdrawn. Publication of such an error in a widely circulated and respected journal like *Nature* would have been a major additional embarrassment in an already embarrassing controversy.

Although passions were clearly running deep over the attacks on Fitch and Miller's date for the KBS Tuff, and were further exacerbated by Curtis' intervention, the reception of his formal presentation was apparently cool, just as had been planned. "Professor Curtis and his colleagues have done a really fine job—with the limited amount of equipment at their disposal," commented Jack Miller. "Jack is a master at damning with faint praise," says Curtis. "Everyone knew he was simply dismissing our results as irrelevant."[3]

Pressure against the Fitch/Miller date was building from other directions, too, including strong statements at the Geological Society meeting by Clark Howell and Frank Brown on the mismatch of the fossils between the Omo and Koobi Fora. And, of course, Basil Cooke once again argued that the fossil pig data indicated that the controversial tuff simply had to be close to 2 million years old, not 2.6. His assertions hadn't changed from the previous presentations, but the quality of data on which he made them was much improved.

The assault on the radiometric date of 2.61 million years was therefore more comprehensive and more threatening than ever before.

Lined up in its defense was an impressive paper by Fitch and Miller, who presented a disquisition on the elegance and superiority of argon-40/argon-39 dating, which they had used on the KBS material, over the conventional potassium/argon technique employed by Curtis. They also cast doubt on the radiometric time scale that Frank Brown had developed for the Omo, which, if correct, would invalidate comparisons between the two sites. Richard Leakey said that the lists of fossil species from Koobi Fora were being expanded and revised and in all probability would show the 2.6-million-year date to be correct. Andrew Brock argued that paleomagnetic-reversal data were consistent with Fitch and Miller's data, though the picture appeared to be rather more complex than initially envisaged. And Kay Behrensmeyer and John Harris, in separate presentations, promoted the merits of the ecological hypothesis. (Harris' paper, which was coauthored by Leakey, was not published in the conference volume, because not long after the meeting they came to realize that the hypothesis was simply untenable.)

So the battle lines were clearly drawn, and the atmosphere became quite tense. At one point Basil Cooke, attempting to lighten the proceedings somewhat, pointed to his "MCP" tie and said, "You might think this stands for 'male chauvinist pig,' but in fact it means '*Mesochoerus* correlates perfectly.' " *Mesochoerus* is the term for a type of fossil pig. His quip provoked a good deal of laughter, but many people were simply not amused.

Leakey, for instance, was furious. "I was prickly and upset," he now says, "because we had some beautiful pig fossils from Koobi Fora and they had been subverted into mere data in a story of chronology that didn't suit my ideas at all."[4] Leakey berated Cooke for not having told him what he planned to say in his paper. "I felt that Basil, as a member of 'my' team, should not have used the Koobi Fora data in the way he did without giving me a full report before the meeting."[5] In explanation of his actions, Cooke now says, "I did not give Richard any report before the Geological Society meeting because I assumed he already knew my views from 1973."[6] Leakey acknowledges he was in error: "I was jolly angry at the time, but I now realize that I was being immature and silly."[7]

What Cooke did not know at the time of the Geological Society meeting was that he was not alone in his work on the Koobi Fora pigs. Richard had quietly nursed an uneasiness about Cooke's work because of some scientific disagreements that Cooke had had with Louis Leakey over the interpretation of the pig fossils from Olduvai some years earlier. And because Richard had been unhappy with Cooke's conclusions on Koobi Fora from the beginning, he decided to ask John Harris to include a study of the Koobi

Fora pigs in his work on the other fauna from the site. This was early in 1974. Tim White joined with Harris in collecting fossils, including pigs, during the 1974 season. "By February 1975," says Harris, "we had a good sample of pigs from Koobi Fora, which we thought supported Fitch and Miller's radiometric date."[8] This was an important factor in bolstering Leakey's confidence for the February meeting, thus allowing him to continue to discount what Cooke and others were saying.

There was some discussion at the Geological Society meeting that Harris might join Cooke in his study of the pigs, to which Cooke agreed. However, the proposed collaboration never occurred. Amid some confusion and apparent misunderstandings about arrangements for his research trip to Nairobi, Cooke's involvement simply fizzled out. The key to the developments, however, is contained in a letter from Leakey to Cooke on 2 July 1975: "John has made significant progress with his suid [pig] research and I am increasingly inclined towards having him produce his own views." To put it bluntly, Harris' answers on the pigs were more congenial to Leakey than were Cooke's.

Not unnaturally, Cooke was extremely upset by this turn of events, as he had already invested a great deal of effort in the project and had even completed a substantial part of what would have been an important monograph on the fossils. His distress at the growing bitterness of events was such that he decided not to include his paper in the symposium volume—a gesture of defiance he now acknowledges to have been a mistake.

Meanwhile, important developments were taking place in a new aspect of radiometric dating. Fitch and Miller had concluded their paper for the Geological Society meeting with the following proposition: "We believe that resolution of the fundamental difference of opinion regarding the true age of the KBS Tuff ... will not be obtained from further potassium-argon dating of the few samples available: it requires the application of an independent radioisotopic dating technique to the problem."[9] Fitch and Miller had an ace up their collective sleeve, which was in the form of fission-track dating.

Anthony Hurford, a student of Fitch's in London, had for two years been working on the application of fission-track dating to the volcanic material from Koobi Fora. He had been joined in the enterprise late in 1974 by Andrew Gleadow, a young researcher from the University of Melbourne. At the time of the Geological Society meeting, Hurford and Gleadow were poised to obtain their first results on the KBS Tuff.

Fission-track dating is extremely simple in concept. Many volcanic ashes contain zircons, which are crystals of zirconium silicate. This mineral always contains trace amounts of uranium, including the isotope uranium-238. Being radioactive, atoms of this isotope decay at a precisely known rate. The U-238 atom explodes, and tears its nucleus into two halves, which are propelled with great force in opposite directions through the crystal. As a result, every time an atom of uranium-238 undergoes fission, a tiny tunnel is bored out through the crystal lattice. The principle of fission-track dating, therefore, is simply to count the number of these fission tracks contained in a zircon crystal, which gives a measure of how old the crystal is: the more tracks there are, the older is the crystal. As with potassium/argon dating, the formation of the crystal in a volcanic eruption resets the radiometric clock to zero.

In 1974, when Gleadow and Hurford joined forces on fission-track dating, the scientific literature of the time seemed to imply that the technique, simple in concept, was also simple in practice. But, remembers Gleadow, "It most definitely was not."[10] The most horrendous part was in etching out, or enlarging, the tracks, which are minute when initially formed. The tracks need to be enlarged so as to be visible for counting under a microscope. Etching the tracks involved steeping the crystals for several days in a molten mixture of virtually 100-percent sodium and potassium hydroxides, one of the most corrosive and unpleasant chemical concoctions one can imagine. A typical result would be that the crystals, which had to be mounted in transparent Teflon for ease of handling, would be lost. Eventually, these and other problems were cracked, and the two young researchers began to etch their first KBS Tuff zircons on 7 February 1975, just two weeks before the Geological Society meeting. They started counting the tracks in mid April, and would soon have an estimate of the date.

On 15 April, Garniss Curtis wrote to Frank Fitch telling him of his further work on conventional potassium/argon dating. The results were the same as he had presented in London: 1.8 million. He then went on to refer to the promise of the independent, fission-track dating that Fitch had alluded to at the Geological Society meeting. "No word from you or Hurford on your fission-track results, hah!" he said with triumph. "It seems likely to me that you have confirmed our results using fission-tracks on the zircons, right? If so, you are preparing a note to *Nature*, right? We have also written up our results. . . . May I suggest that we put out back-to-back papers?"

Right? Wrong.

"My first note in my lab book on the results of our counting was on 9 May,"[11] recalls Hurford. "Ten crystals counted, with a mean age of 2.4 million."

On the following day Fitch wrote a brief letter to Leakey. "Tony Hurford and Andy Gleadow are working independently on the fission track dating of the zircons from East Rudolf. They gave me a progress report yesterday. The mean fission track age is 2.62 ± 0.40. The work continues." Here, apparently, was a stunning confirmation of the original Fitch/Miller date of 2.61 ± 0.26 million. Leakey, naturally, was delighted, for it seemed that his obdurate stance at the February meeting had been justified. "Could you please ask them to keep up the good work!"[12] he replied. They did, and a month later Fitch felt confident enough to assure Leakey that "It is now certain that fission track dates will be able to provide an independent yardstick for the East African Plio/Pleistocene successions."[13]

Fitch eventually did reply to Curtis' letter of 15 April, saying that the two labs were no nearer agreement. "Andy Gleadow and Tony Hurford have been working *independently* on zircons separated from KBS pumice lumps collected at the site below 'the headland' in Area 131 (near where I believe Thure [Cerling] to have collected his sample). The mean age of the first 14 crystals analysed is 2.62 ± 0.4 m.y."[14] These findings impressed Curtis. "While the Fitch and Miller results are easily explained," he later wrote to Leakey, "the fission-track results are puzzling."[15]

Nevertheless, Curtis and his colleagues prepared a manuscript of their initial findings, with the intention of submitting it to *Nature*. A copy reached Leakey's hands, via Glynn Isaac. Leakey immediately penned the first of a pair of strongly worded letters to Curtis. "In the main part, I am glad to see the work advance and if the dates you produce are the real age of the KBS Tuff(s), nobody could be happier than I to see this problem eventually settled," he began his letter of 3 June. "Nonetheless, I have grave doubts still and I would like to see further dialogue before publication." He then went on to say that he was disappointed with Curtis' handling of the affair. "You appear to have been aware that I initiated the work at East Rudolf and that I was in overall command of the programme. In spite of this, I have not had a single letter from you about the dating programme. I hear rumors of your results, I hear suggestions, conflicts, etc., but never once a straight forward letter from you. You have dealt through students, colleagues and members of my family—I feel insulted by your actions."

Leakey considered the inferences Curtis and his colleague drew in their manuscript concerning the relationship of the fossils between the Omo and Koobi Fora to be invalid. He therefore suggested that if they wanted to go ahead with the paper, it should be accompanied by a rebuttal on these issues by Fitch, Miller, Harris, and himself. "I am deeply concerned that we do not further confuse the consumer by presenting another partial statement that could well prove to be incorrect."

A few days later, Leakey wrote to Fitch that "There seems to be a desperate attempt to undermine our combined research endeavours and this must not succeed."[16]

Curtis' 15 July response to Leakey's outburst was to go over some of the arguments as to why he considered his potassium/argon date to be correct and Fitch and Miller's to be flawed. He also enclosed a copy of the final draft of his manuscript to *Nature*, saying, "We had hoped that Fitch and Miller would accept our suggestion to join us in publishing back-to-back articles, theirs presenting the new fission-track data. Apparently they are not interested in this idea." In fact, although Miller had a strong interest in the outcome of the fission-track dating, he was never directly involved in the work. A day later Curtis mailed the manuscript to the *Nature* offices in London, where it was received on 21 July.

Leakey replied immediately with a second emotional letter. "I would have preferred a more objective approach to the joint programme and I cannot accept that you have helped towards this goal. Frankly, I cannot help but regret your submission of the present article and this is not because I don't like your numbers; they are of little consequence to my own appreciation of the East Rudolf collections. I merely think your paper is misleadingly subjective, incomplete and premature."[17]

This time Curtis responded with an even more extensive explanation of his own position, detailing step by step how and why he had got involved and why Fitch and Miller's experimental results and interpretations were suspect. But he began by saying that "I won't deny your charges of unobjectivity. I'm married to a clinical psychologist who constantly points out to me how unobjective scientists are in general and unobjective I am in particular; however, I believe that I have *tried* to be objective in our paper concerning the age of the KBS Tuff."[18]

Curtis explained that even without knowing of Dalrymple and McDougall's critical reviews of Fitch and Miller's Nairobi conference paper, he was extremely disturbed by what he read in it. "Their choice of 2.6+ for the age of the KBS Tuff seemed to be by their reaching into a hat filled with all the numbers they had obtained and coming out with 2.6 m.y." Their measurements of argon produced radioisotopically in the crystals showed very heavy contamination with argon from the atmosphere, he said, which implied that large errors might be included in their interpretations. He then went on to explain that in his initial potassium/argon measurements he had obtained 1.85 million on two types of crystal from the KBS Tuff material: glass and sanidine. If the tuff had been heated, as the overprinting idea demanded, then glass and sanidine would have given different dates in the experiments, because glass fragments alter faster under such conditions. "I thought that pair of dates would shake up Fitch and Miller and that they would finally understand that they had a much bigger problem than they thought," he wrote. "My lecture in London, which was incomprehensible to most of the audience, was meant for Fitch and Miller. They failed to see its application to them and gave it condescending praise."

Leakey did not reply, perhaps because a few weeks earlier he had received emphatic assurance from Fitch that all was well with the KBS. "There is still no scientific reason for doubting the 2.6 m.y. age obtained from the KBS Tuff," Fitch had written on 16 June. "In contradistinction to this careful work we did on the KBS, anyone who listened to and understood what I said at the February Conference must know that there is every reason for suspecting the accuracy of unsupported conventional degassing potassium-argon dates (especially those from young rocks). . . . I am afraid the present efforts of Garniss and his associates to date such rocks by simple conventional potassium-argon methods are just irrelevant." He went on to say that Curtis' techniques were "too inaccurate and old-fashioned to be acceptable." He conceded that there were no easy answers, and said, "They will certainly not be obtained by primitive analyses undertaken on a few samples of doubtful provenance by geochronologists who have not visited and studied the field locations."

However, Leakey was in for a shock. Glynn Isaac, apparently more and more impressed with the gathering weight of evidence, in particular Curtis' potassium/argon date which Isaac had initiated, changed his position. He could no longer support the Fitch/ Miller date, he decided. "Glynn has totally abandoned the ship," Leakey wrote to a close colleague in June 1975. In spite of this change of mind, Isaac continued publicly to support Leakey in his championing of the Fitch/Miller date, largely because of his strong sense of loyalty to his colleague and co-leader of the Koobi Fora project. By the fall Richard himself had begun to waver, strongly influenced as he was by Isaac's change of mind. On 8 October, Mary Leakey wrote to Garniss Curtis, "Of course, I never can discuss this with Richard, but I hear on the grapevine that he thinks you are probably right." Mary Leakey maintained a regular correspondence with Curtis throughout the whole of the KBS Tuff controversy; the two had been close friends ever since Curtis carried out dating work for Louis and Mary in the early 1960s, and Curtis was now dating Mary's new site at Laetoli.

From this point—late 1975—onward, the solid Fitch/Miller/ Leakey phalanx of support for the 2.6-million-year date came under ever-more-vigorous attack, and eventually began to crumble. Fitch and Miller continued to defend their results, and in a sense still do; whereas Leakey began to be more and more openminded about the problem. He was, however, unable to bring himself to accept Curtis' results at face value. In fact, in June 1976 he wrote to Thure Cerling, "I do not think it useful or wise for Garniss to do any further dating of our material for the moment."[19] Instead, he eventually became convinced by the work of the independent potassium/argon geochronologist Ian McDougall. But along the way, Leakey's assessment of the KBS Tuff problem was buffeted by two separate lines of evidence, which sometimes seemed to imply that Fitch and Miller's 2.61-million-year date was right and sometimes that it was wrong.. The first was John Harris and Tim White's final conclusions on the fossil pigs; and the second was further work on fission-track dating.

The process by which Leakey came to realize that he had been mistaken was, therefore, by no means rapid and clear-cut, with scales dropping from the eyes. Rather it was a slow, halting affair, with many moments when it seemed that the older date might be salvaged after all. One reason it was slow was that, having championed the older date for so long and with such passion, Leakey would scrutinize each new item of evidence for anything that might be construed as support for his original belief. "I was never in a position properly to judge whether or not Frank and Jack's date was right, because I knew very little about geophysics and dating," Leakey now concedes. "Instead I supported them vigorously, because that's what I felt I should do. I got myself so deeply wedded to their date that I think I lost the ability to assess the evidence truly objectively."[20]

It is also the case, however, that to some extent the evidence that was being produced was itself subjectively influenced, and tended, in the case of the fission-track dating at least, to give the answer that was expected of it rather than the one that was objectively correct. Leakey therefore had plenty of opportunity to continue to believe in the older date, which is what he wanted to be able to do. Eventually, objectivity did force its way into the controversy, but not before a great deal of anguished to-ing and fro-ing on the parts of the various protagonists. The denouement of the KBS Tuff controversy therefore illustrates not only that it is possible to be wrong in science, even with the apparently straightforward task of obtaining a single date for a single volcanic tuff; but also that typically there is a degree of uncertainty in science that is not often made public, because it is contrary to the mythology of what science is supposed to be like.

When Harris and White first examined the pig fossils, in 1974, their initial impression was that a 2.61-million-year date was correct for the KBS Tuff. They were getting the "right" answer, which, White now says, he and Harris were more than happy to get. "We were caught up in the atmosphere of the whole thing, the mythology about Koobi Fora and Richard Leakey," recalls White. But they soon faced problems with the geology at Koobi Fora. "The more we collected, the more we realized that the stratigraphic program we were supposed to operate in just didn't work. Nothing fit."[21]

The dictionary definition of stratigraphy is the "order and relative position of strata." In the field it means being able to recognize the same layers of rock in different localities. It is often a very difficult task, especially at Koobi Fora. Looking in some places like a moonscape and in others like a bulldozed building site, the Koobi Fora sediments offer a nightmarish challenge to the geologist who wishes to make accurate correlations of strata across the full extent of the 2,000 or so square kilometers that encompass the fossil-collecting localities. In this case, the nightmare came true: the stratigraphy that had been worked out by 1975 was a mess.

"It became clear that the correlations we were making between the three main areas on the east side of the lake were not always accurate,"[22] Leakey now says. "For instance, the type site for the Tulu Bor Tuff is up north, at Ileret. It's very thick, about eight or nine feet. And below it you can see the Surgei Tuff, which is the bottom of the succession. People assumed that when they found the bottom tuff in other areas, in Koobi Fora and Kubi Algi, that it too was the Surgei, that the one above it was the Tulu Bor, and so on. But it turned out that we were calling different tuffs in different areas by the same name. It was a mess." The uncertainty over the dating of a single, albeit important, tuff was bad enough. But add to this a thoroughly confused stratigraphy, and one can imagine what a daunting problem it all seemed. Bill Bishop's warning about the difference in the number of recognized tuffs between the Omo and Koobi Fora, which he delivered at the end of the September 1973 meeting in Nairobi, had come to fruition. "I virtually lost hope that we would ever solve these problems of stratigraphy and dating with the people we had,"[23] recalls Leakey. "And it was becoming increasingly difficult to get funding."

As the pig fossils had served to highlight the problems of the stratigraphy, it was appropriate that Harris and White should be at the center of the initial attempt to salvage the situation.

The two young researchers were able to demonstrate to Leakey just how bad the stratigraphy at Koobi Fora really was, primarily because during the previous eight months they had finally come to understand what the pig fossils really meant. Having discerned the patterns that were clear for the pigs from other parts of East Africa, they could see that the stratigraphy as then described by the Koobi Fora geologists just had to be wrong. They could also see that the KBS Tuff had to be younger than Leakey believed it to be.

This realization began when White spent a month in Berkeley early in 1976 measuring and photographing the pig fossils in Clark Howell's charge, which were from the Omo. This was the first time White had had an opportunity to study in any detail pig fossils from elsewhere than Koobi Fora, and the experience proved to be crucial. A picture began to form in his mind, but he wouldn't be certain about it until he'd seen the rest of the Omo pig fossils in Paris, where they were in the charge of Yves Coppens. White arrived in Paris on 19 May and made his way to a dungeonlike laboratory at Chatenay, just outside the city, where the fossils were stored. He again measured and compared the fossil pig teeth, and knew that his suspicions had been correct. John Harris joined White on 25 May, and the two worked together before traveling on to Nairobi a week later. This was the first time Harris had seen any of the Omo fossils, but he too quickly began to discern a clear picture. "What are we going to do?"[24] he asked White. "John, there's no way around this." "People aren't going to like it," Harris said. "No," replied White, "but we started this study and we are going to publish it."

"We had seen that the correlation that Basil Cooke had made had been correct," explains White. The pigs under the KBS Tuff couldn't be 2.6 million years old, as they had once thought, but instead were more like 2 million. Cooke's comment on their conclusions, after having been effectively edged out of the project, is poignant: "My only real consolation was that their study, commissioned to refute my dating contentions, in fact only served to confirm them."[25]

When Harris and White reached Nairobi at the beginning of June after their study session in Paris, Harris told Leakey what they had discovered. "He accepted it almost without a murmur,"[26] says Harris. This, however, turned out to be the proverbial calm before the storm. During the two months after they arrived in Nairobi, Harris and White spent much of their time further analyzing and writing up the pig data for an important monograph on the subject. Then, in the middle of August, they joined Leakey and several of the team's paleontologists and geologists for what turned out to be a lively but often rather bitter meeting up at Koobi Fora. Leakey had called the meeting so that the problem of the stratigraphic mess could be faced straight on. Everyone anticipated that it might be a difficult time, and it was.

Once they were at Koobi Fora, Harris and White drove from site to site, pointing out the problems as they saw them. Kay Behrensmeyer and fellow geologist Ian Findlater explained the stratigraphy as they saw it. No one was satisfied. "Ian was unhappy about having his conclusions questioned," explains Harris. "And we were very unhappy at having to fit what seemed like a logical evolutionary sequence into a scheme that made nonsense of it."

Eventually it was decided to give up any attempt to correlate between the three main areas and instead to impose three separate tuff-numbering systems, one for each of the three major areas. It was a complex affair that few people outside the Koobi Fora project ever understood. Findlater refused to have anything to do with it. "It was meant to be an interim arrangement that would allow us to place fossils in relation to known tuffs within an area without trying to correlate between areas,"[27] explains Leakey.

At the end of it all, Leakey said to Harris and White, "OK, you two, you can go back to Nairobi now and write up your pigs. You've done enough damage here."[28] The two set off the next day on the rugged forty-eight-hour drive back to Nairobi and immediately began to prepare a manuscript for *Nature*, essentially by abstracting material from their recently completed monograph. The *Nature* manuscript was soon to be the cause of another storm.

"Finished *Nature* paper, John typing it . . . tonite typed final tables," recorded White in his daily journal on 3 September. "All that's left is the photos." With the final details tied up, Harris then took the manuscript to Leakey, who had recently returned from Koobi Fora in his single-engine Cessna, which does the journey in two hours forty minutes. This was just a couple of days before White was due to leave for an international conference in Nice, where he was expecting to stand in for Leakey and present a short report on a recently discovered *Homo erectus* cranium.

"It was a very good account of the pigs,"[29] Leakey now says of the manuscript, "but at the end there were a number of pages on the implications for hominid chronology, and this included discussion of a number of specimens that hadn't yet been published." This latter proved to be a red flag.

Harris and White had written the paper together, but Harris was principally responsible for the pig section, while White had done the discussion on the hominids. "We were putting it together, and it was all very exciting,"[30] says Harris. "So we followed it through to the logical conclusion." The logical conclusion was simply to say that the KBS Tuff must be closer to 2 million years old, not 2.6, with the obvious implication for the hominid fossils recovered from below it—including, of course, 1470. The skull 1470, remember, was held to be the oldest member of the genus *Homo* yet discovered, and was therefore the jewel in the paleoanthropological crown of Koobi Fora—that is, if it was truly at least 2.6 million years old. If, as Harris and White's manuscript implied, 1470 was younger—more than half a million years younger—then the significance of Koobi Fora would inevitably be dulled somewhat.

Leakey was furious with what he saw in Harris and White's manuscript. Not, he insists, because of the implications for the hominid dating, but because the manuscript violated a long-standing group rule. "The rule says that members of our group must not discuss in print hominid fossils that have not yet been published in the *American Journal of Physical Anthropology* by the principal investigator working on the fossils,"[31] says Leakey. "The rule was instituted to protect the interests of those people who did all the donkeywork in formally describing the fossils. Tim and John's manuscript included some fossils in this category."

Leakey therefore told Harris in no uncertain terms that inclusion of hominid data in the form submitted was quite unacceptable. The exchange between Leakey and White was brief, and is now subject to conflicting accounts. Leakey describes it as an explosive row, with him telling White that he had broken project rules over the hominids, and that if he went ahead with the manuscript as it was, he would be out of the project. White is said to have accused Leakey of scientific censorship, and to have rushed out of the office, slamming the door as he went. By contrast, White claims that Leakey merely said he wasn't in agreement with the paper but that a few changes would be all that was necessary to make it acceptable. White then went to Harris' office, and only then learned of the extent of Leakey's objections. "When I found out . . . I lost my temper," he recorded in his journal. "I was really shaken up—actually crying because of the insult."

In any case, White was in no humor to discuss anything with anyone, and he quickly left the Nairobi museum in company with Harris, later to drive to the airport for the flight to Nice. "Tim was extremely angry,"[32] recalls Harris. "He said he wanted nothing more to do with the paper and nothing more to do with Kenya while Richard was in charge. It was as a result of this incident that Tim developed the idea that Richard was trying to suppress scientific information." Harris was always in a difficult position throughout this episode. Not only was he head of paleontology in the Louis Leakey Institute attached to the museum, but he was also Leakey's brother-in-law.

After a few days in Nice, White cooled down sufficiently to write a long letter to Leakey explaining his reaction and his thoughts. He argued that all the hominid fossils mentioned in the manuscript had been published in *Nature* and therefore described to a degree sufficient to allow the kind of mention he and Harris had given them. "I have very strong personal feeling about this, not because I wrote the section, but because the material is being given special status and hidden from view and denied even the most restrained discussion. . . . *Anyone* who had read your own *Nature* and *American Scientist* articles could have written the same thing as myself. Thus, I have not held a privileged position, only commented on fossil material to a level that is available to any responsible reader of *Nature*. If scientists cannot do this I no longer wish to take part."[33] White's disposition was not improved when Glynn Isaac approached him and urged a "more diplomatic" approach, avoiding the inclusion in the paper of numbers that might offend "some people." White interpreted this as further evidence of scientific censorship, and would have none of it.

Leakey responded to White by repeating the group rules and adding that he and Harris should have circulated their manuscript for discussion and comment. Instead, wrote Leakey, "I was given a photocopy of a completed version for filing. This is both wrong and disrespectful."[34]

It is perfectly true, as White claimed, that any scientist could legitimately have written about the hominids in the way that he and Harris did, simply using the material already available in *Nature*. Being part of the Koobi Fora project, given its strict rules of operation, was therefore, in terms of freedom of expression, a disadvantage rather than a privilege. White, who has something of a passionate and sharp-edged personality, ran smack into it and exploded. Later, White would write to Leakey, "Last year's problems stemmed from my ignorance about policy and I want to make sure I don't upset anyone again."[35] But the relationship never really recovered. White continues to claim that the incident illustrates that Leakey chose to make use of the group rules to suppress "uncongenial" scientific information. Leakey insists that it was a matter simply of trying to maintain long-established rules for the good of the whole group.

Harris confesses that he was surprised by the severity of Leakey's reaction to the manuscript. "I was taken aback that he felt unhappy about our mentioning the hominids at all."[36] Harris modified the manuscript, mainly by trimming the hominid section a little, and submitted it to *Nature*. Surprisingly, the journal rejected the paper, saying that one of the referees had recommended rejection because all the information had already been published by Basil Cooke. White says he now suspects that the *Nature* editors had been influenced to reject the paper, "given what was going on at the time."[37] But no one has ever produced any evidence of inappropriate intervention in this case with the journal.

The paper was eventually published by *Science*, which is the American equivalent of *Nature*, on 7 October 1977. Although the paper was virtually devoid of "offensive" dates, its implications were clear, albeit couched in suitably ambiguous language. Although Leakey knew that Harris and White believed that their pig data implied a younger date for the KBS Tuff, he still entertained the notion that their implications might be wrong. The reason was not that he thought the pattern of pig fossils they described might be mistaken, but that the dating sequence used for comparison at the Omo might turn out to be incorrect. It was a thin straw, but one that allowed a continued belief in Fitch and Miller's 2.61-million-year date. And it so happened that this thin straw was given extra substance by the second main line of new evidence that was coming out at this point: that is, fission-track dating.

At the time that Harris and White had given Leakey their original manuscript—September 1976—Leakey had just recently learned of Tony Hurford and Andrew Gleadow's latest results on fission-track dating. The two young researchers had been refining their technique and were about to have their first major results published in *Nature*. The date for the KBS Tuff they eventually came up with, which was established in the 28 October 1976 issue of the journal, was 2.44 ± 0.08 million years—which has about it a clear ring of precision. Yes, this was a little younger than Fitch and Miller's 2.61-million-year date. But no matter, because in the same issue of *Nature* Frank Fitch, Paul Hooker, and Jack Miller published an accompanying paper that presented a recomputed date for the KBS Tuff, based on a revised value of an experimentally generated constant relating to radioactive decay. Their answer was 2.42 ± 0.01 million years. So not only was the original estimate confirmed by the new technique of fission-track dating, but the two dates were again closely concordant. Not surprising, then, that Leakey felt he could interpret Harris and White's pig data as not necessarily a threat to the older date for the KBS Tuff.

Several observers are still uncertain about Fitch and Miller's revision of the 2.61-million-year date to 2.42, which so unequivocally maintained the concordance between the fission track and argon-40/argon-39 dates. "The revision in 1976 by Fitch and Miller of their preferred interpreted age from 2.6 to 2.4 million years did not enhance their reputation as good experimentalists, to put it mildly,"[38] says Ian McDougall. "I still have no idea as to whether the revision was valid." Frank Brown's opinion is similar. "I didn't receive that well at all,"[39] he notes. "There were a stack of things wrong with that paper. I reviewed the paper in fair detail, pointing out things I thought were wrong, but it was published virtually unchanged."

In spite of all this, Fitch and his colleagues took the opportunity to lay out once again all the arguments as to why their argon-40/argon-39 results were right and why Curtis' conventional potassium/argon dates were wrong. It is worth noting, incidentally, that the degree of repetition in this paper is not substantially less than that in Harris and White's manuscript, which was rejected by *Nature* because of alleged repetition of Cooke's data, and may even have been more extensive. In any case, Fitch's paper was essentially a response to the results that Curtis and his colleagues had presented at the Geological Society meeting and had published in the 4 December 1975 issue of *Nature*. "A small programme of conventional total fusion potassium-argon age determinations"[40] is how the Fitch paper refers to Curtis' work. "That's Frank being waggish,"[41] says Miller. Waggish or not, it put into the public arena a sanitized version of the opinion expressed privately but generally by Fitch and Miller about the relevance of Curtis and his "old-fashioned" techniques.

Fitch and Miller also claimed that Curtis and his colleagues were mistaken in believing that two separate tuffs—measured by Curtis at 1.6 and 1.8 million years—were erroneously being called the KBS Tuff. As it turned out, there are two distinct tuffs very close together at this point, but Curtis' separate numbers derived from the use of a weighing balance that had temporarily been out of adjustment, and not from astute geochronology. The inaccurate numbers obtained from the defective balance during this period upset the calculations for one particular batch of tuff material, thus giving the younger date of 1.6 million years. Not unnaturally, when news of this error became public it was received in the Fitch/Miller laboratories as confirmation that their opinion of Curtis had been correct. "It was very unfortunate,"[42] comments Frank Brown. "It gave the other side ammunition, a reason to ignore

Curtis' results." Harris agrees: "Because of that mistake we weren't prepared to give much credence to Curtis' work."[43]

The publication of the fission-track results had not been at all straightforward, and the background to it illustrates the kind of pressure under which the young researchers had been working.

Gleadow had left Fitch's lab in October 1975 and had gone back to Australia via the United States, where he spent some time with Charles Naeser at the U.S. Geological Survey in Denver and with Garniss Curtis and his colleagues in Berkeley. He had left, he thought, with an oral agreement all around that the fission-track results obtained to that point were premature and would not be published. He was therefore more than a little surprised the following March to receive a draft of a paper with his name on it, which presented the initial results for the KBS Tuff. Fitch had been extremely enthusiastic about getting the new data published, Hurford remembers, and he therefore "encouraged" his young student to put together a manuscript. "At first I was extremely unhappy about publishing anything at this stage because we'd all agreed not to and for very good reasons,"[44] Gleadow wrote to Hurford on 17 March 1976. "However, two days ago I got a copy of Chuck [Naeser]'s letter from you with his determinations on two more KBS and one Karari sample. That was encouraging to see and it now seems absolutely clear that no matter what we do we cannot escape from an age of ca. 2.4 m.y. for the KBS."

Gleadow, a cautious and thorough worker, had been uneasy about his and Hurford's initial results, primarily because the technique was in its infancy and the samples counted were relatively few in number. However, the fact that Naeser had, apparently independently, worked on similar material and had come up with the same result served to sweep aside his caution. Here was another example of the cogency of independent replication. Gleadow therefore agreed to go ahead with the *Nature* manuscript, with Naeser as third author. He wrote to Leakey at the same time, saying that "After a critical assessment of all the data now in hand, I am convinced that a fission track of 2.4 m.y. for the KBS Tuff . . . is quite inescapable."[45] Leakey had been looking to fission-track dating as a way of resolving the conflict between Curtis and Fitch and Miller, and the answer he was given could not have been more unequivocal. This was just a few months before Harris told Leakey about his and White's conclusions on the pig fossils.

Having agreed that the *Nature* paper should go ahead, Gleadow felt uncertainty begin to stir in his mind, in particular about the methodology that he and Hurford had developed. Was it as objective and free of bias as they had imagined? By November he was full of doubt and concern. He sat down and typed a long letter to Hurford, explaining in detail his reservations about the validity of their work. "I have grave doubts about the validity of the 2.4 m.y. ages for the KBS zircons."[46] he concluded. Hurford remembers that this was the only letter to him that Gleadow had typed, rather than written in longhand. "When I opened it, I knew immediately that something was wrong,"[47] he remembers. After waiting a few days before replying, "to cool off," Hurford answered Gleadow at some length, conceding that "a remote possibility exists with such low track density material that *without knowing it* unintentional bias to influence the results either to 1.8 or 2.4 may be creeping in, either because of outside pressure (more likely in my case) or because of acceptance of an hypothesis in our minds."[48]

Hurford recalls that Frank Fitch and Jack Miller thought that Gleadow must have been "got at" when he was in Berkeley, hence his apparent defection. "It's true that talking with Garniss and Bob Drake I came to see that there was another legitimate side to the argument,"[49] says Gleadow. "We had been sheltered from that in London. I realized then that their viewpoint had to be taken a lot more seriously than was happening at that time in London and Cambridge."[50]

Gleadow wrote to Leakey at the end of 1976, shortly after his typed letter to Hurford, alerting him to his concerns: "I now have strong reasons for doubting the reliability of our earlier fission track measurements for the KBS Tuff and think that the apparent ages of about 2.4 m.y. should come down."[51] A better estimate, he suggested, would be just less than 2 million years. "My own reaction is to maintain a low profile and to continue the investigations so that a final solution can be eventually reached,"[52] Leakey responded. "The research group is now able to deal rationally with the dating issues and this is itself a great relief." Leakey now recalls that "When I heard about Andy Gleadow's reservations about the fission-track dating, I knew we were in trouble."[53]

But Gleadow's position was by no means settled. "I swung from having serious doubts about the 2.4 date to being equally certain that it was correct,"[54] he recalls. For instance, less than a year after these exchanges with Hurford and Leakey, he once again was supporting an age of 2.4 million years. On 10 August 1977 he sent Hurford two results on the KBS "that I am happy with." They were 2.42 and 2.30 million years. Hurford, he said, was free to use them at the upcoming Pan African Congress on Prehistory which was to be held in Nairobi. Leakey therefore again had good reason to hope that the older date might be correct after all.

But Gleadow was in turmoil, swinging between certainty and doubt. By the end of 1977 he was again seriously questioning the 2.4-million-year date. Then in February 1978 he again wrote to Hurford with yet more new results: "they give 2.4 m.y. all the way."[55] With uncertainty of this sort in the mind of an extremely able scientist, skilled in a very important but difficult technique, it is perhaps not surprising that Leakey's recognition that the older date for the KBS Tuff was indeed wrong would at best be halting. Fitch and Miller remained unswerving in their support of the older date.

What was causing Gleadow concern was the ability to discriminate genuine as against pseudo tracks in the crystals, the reliability of counting them, and the regimen for plugging the numbers into the age computations. It was a whole series of things, each small in itself, but in combination potentially significant. He worked on each component that worried him, and finally, in July 1978, he got everything together and counted a series of crystals using what he believed to be, at last, a bias-free procedure. "I did a month's counting without doing calculations,"[56] he explains. "Then I did the calculations. I stayed up till two in the morning. The numbers started to come out: 1.8 . . . 1.8 . . . 1.8. I could hardly believe it."

A few days later Gleadow left Australia for the United States, where he would meet Hurford on a geological field trip in Wyoming. The trip had been organized by Carl Vondra, of Iowa State University, who had been involved in geological work at Koobi Fora in the early 1970s. Frank Fitch was to be on the trip too. The gathering was a prelude to an international conference on Geochronology, Cosmochemistry, and Isotope Geology, to be held at the end of August in Snowmass-at-Aspen, in Colorado.

Not surprisingly, Gleadow was a little apprehensive about how he would break the news of his redating to Hurford, but more especially to Fitch. He told Hurford first. "I was in a state of bewilderment,"[57] recalls Hurford. "We talked it over a lot. I was convinced his analytical approach was right. But I wasn't convinced that 1.8 was right. I had nailed my colors to the mast of 2.4 and I didn't want to believe 1.8." Vondra helped break the news to Fitch, simply by bringing up the subject over dinner one evening. "Frank was very polite about it—made a few comments, but there was no big scene or anything,"[58] says Gleadow. "That was it."

Hurford was scheduled to present a paper at the Snowmass meeting, with Fitch as a coauthor, on the "unresolved controversy" of the KBS Tuff. "After hearing about Andy's results I refused to give it,"[59] says Hurford. "Frank had to give it instead. He was very angry about that. He said something about new fission-track results introducing some uncertainty, that's all."

In looking back over events, Gleadow and Hurford now realize there were several factors that led them astray. For instance, says Gleadow, "It was never true that Tony and I were doing the work independently of each other. We developed the techniques together, we looked down the microscope together, we agreed what were tracks and what weren't, together."[60] The same applied to Naeser. "We worked so closely together, all three of us, that it was in no sense independent." Gleadow later acknowledged this publicly when, in March 1980, he published his new results in *Nature*. "Note that the zircon fission track ages of Hurford *et al.* represent the first attempt at dating young zircons of this kind. These ages apparently agree closely with each other but this is mainly due to the close communication between these authors on track identification and discrimination in these samples."[61]

Hurford now identifies as an important factor their practice of counting tracks and calculating the age crystal by crystal. "You can bias your results 10 percent either way, easily,"[62] he says. "You go crystal by crystal, and you begin to see where the rolling average is going. If you need the count to be higher with the crystal you're working on, so that it will fit in, you might include something that is a doubtful track. If you want the count to be lower, you don't include it. That was poor practice." Gleadow agrees. Even so, there was an undercurrent of uncertainty from the beginning. "I was never confident of our results,"[63] says Gleadow. "But that was overwhelmed by the desire for our numbers. When you are a young post-doc you don't say everything that's going through your mind."

The obvious question to ask, given the lack of precision of Gleadow and Hurford's early fission-track dating attempts, is how was it that the answer they got was so close to what Fitch and Miller would so obviously find highly acceptable? "Frank didn't come in and say, 'No, that's not old enough, I want 2.6,' " says Gleadow. "But if you got a number out of a scatter of numbers in your first crude attempts around 2.5, boy, was that praised—great excitement and so on. It all builds up." Hurford agrees. "Remember, we got our first tentative date of 2.4 on the ninth of May, and the next day Frank is writing to Richard to tell him the good news. That's the kind of pressure there was."[64]

"With hindsight it is clear that we were really very naive about the difficulties of dating geologically young zircons,"[65] Gleadow observes. "However, the highly charged atmosphere surrounding the KBS controversy meant that any result we produced took on an exaggerated significance and produced a certain feeling of implicit pressure (imagined maybe) to publish without delay. I wish we had waited but I guess that is the lesson of the whole KBS story."

The final resolution of the KBS Tuff controversy stemmed from Ian McDougall's intervention with a new and complete set of potassium/argon and argon-40/argon-39 dates. The continuing uncertainty had motivated Leakey and Isaac to invite him to do a dating at Koobi Fora; and although McDougall at first was reluctant, he had eventually agreed, about a year before the Snowmass meeting, to undertake the project. "Ian had a very good reputation," says Leakey, "and by involving him there seemed to be a possibility of resolving this terrible problem. It had dragged on too long and was being detrimental to my professional life and my health. I knew that if there was to be a resolution, it had to come from someone who wasn't involved with either Fitch and Miller or with Curtis. Ian offered that possibility."[66]

"By the time of the Snowmass meeting I had some preliminary results of about 1.9 million years on the KBS material," McDougall now recalls. "And I discussed them with Andy Gleadow at the time."[67] Gleadow therefore knew about McDougall's preliminary date before he did his recalculation of the fission-track date, a fact that must have given him confidence that the line of 1.8 . . . 1.8 . . . 1.8 numbers was in fact correct.

McDougall went to Koobi Fora after the Snowmass meeting, in company with Gleadow, Garniss Curtis, and Robert Drake, and collected more material for a thorough dating program. "No, I wasn't disappointed when Ian told me his results," says Leakey. "By this time I wasn't really surprised that the Fitch and Miller date would be shown to be wrong. The main feeling was one of relief—relief that this whole sorry business was finally over. I was quite ill at the time, in the final stages of kidney failure, and sometimes the KBS problem had become more than I could cope with. Now it was over, and, yes, I was very relieved, despite the outcome."[68]

McDougall published the first results from his extensive dating program at Koobi Fora in the same issue of *Nature* that carried Gleadow's revised fission-track data, in March 1980. According to an editorial by Richard Hay, this duo of papers "may have ended a decade of controversy over the age of the KBS Tuff in the East Turkana region of Kenya."[69]

The question at the end of it all, of course, is Why did it happen? Was it a severe technical problem that would have entrapped anyone who ventured into it? Or is the answer more in the realm of the sociology of science? For instance, did Fitch and Miller stick unreasonably long with their initial date, because, for instance, they had their reputations to uphold? Or did Leakey keep the pressure on his team and on the British scientists, so as to maintain his ancient *Homo*? Or was there something else entirely?

From the strictly technical point of view, both Frank Brown and Ian McDougall are in no doubt that the KBS Tuff presents no special problems for either conventional potassium/argon or argon-40/argon-39 dating. For instance, Brown's assessment is this: "The KBS Tuff has excellent material for dating. I don't understand why they had so much trouble."[70] McDougall's comprehensive dating program throughout Koobi Fora produced consistent, repeatable results, with no complicated age spectra from the argon-40/argon-39 technique that would have required unusual explanations. And this was true for the KBS Tuff just as much as for the others in the sequence. In publishing this comprehensive set of data in 1985, McDougall commented on the KBS history, saying that "the large spread of results reported by Fitch and Miller must result from experimental difficulties or error estimations that do not properly reflect the uncertainties in the actual measurements."[71] Translated from the polite, measured language of the scientific paper, this in plain English means that in McDougall's opinion, Fitch and Miller's experimentation and computation was not up to scratch.

Fitch and Miller absolutely and completely reject this suggestion. They simply dropped the idea of overprinting, which in truth had never been a very convincing explanation to geologists who gave even cursory consideration to the suggestion. Instead, Fitch and Miller now contend, the reason they got such a confusing spread of dates was that the material they were given to analyze had been mistakenly collected from a series of tuffs other than the KBS. "Not all the samples came from the KBS Tuff,"[72] states Miller flatly. And this, of course, includes that first sample site from which the Leakey I feldspar crystals were collected, the ones upon which Fitch and Miller were hung up for so long. Leakey finds this explanation less than convincing. "It is a most extraordinary explanation,"[73] he says. "I am as sure of where that first sample site is as I am about where my house is."

Both Fitch and Miller admit that they clung to the 2.6- (later 2.4-) million-year date because of the exquisite state of those original crystals. "They were beautiful crystals, so we felt confident about them,"[74] says Miller. "Having been given those beautiful crystals, which gave us 2.42, we thought there must be something wrong with the other dates, the 1.9 and so on,"[75] adds Fitch.

Early in 1981, Fitch came across a glass vial containing some of the original crystals—the ones that were the cause of it all—and decided to have them dated by another laboratory. He sent them unidentified and amid a batch of other material to John Mitchell, a former student of Miller's, now at Newcastle University. When he got the answer, he wrote to Leakey: "I remain mystified about the original Leakey I sample from the 'KBS' (?). Recently I have had the remainder of the collection of nearly perfect ⅛" to ¼" crystals sent to us in 1969 dated (as an unknown sample) by an independent laboratory—their answer, using the new constants, was 2.3 Myr! Certainly, Jack's dating was not at fault and, as the sample gave every appearance of homogeneity and freedom from contamination, it would seem that it must have been obtained from a *tuff or pumice blocks* OLDER than the tuff now known as the KBS in Area 131 and elsewhere. Have you any suggestions that might resolve this problem?"[76]

Leakey's response was brief: "The Leakey I sample was collected by Kay and myself at the type locality and of this there is no doubt what so ever. We only collected there. I would be most interested in an answer to this problem myself. . . . Like you, I remain doubly curious if Jack's dating really was not at fault."[77]

So, did the infamous Leakey I crystals come from a tuff other than the KBS? According to McDougall, "There is little doubt that their samples came from the KBS Tuff." The reason, he explains, is that "We have found no evidence for a tuff of age ca. 2.4 Ma at Koobi Fora."[78] Brown agrees: "I view it as extremely unlikely that there is a tuff of this [2.4 m.y.] age at Koobi Fora."[79] If there truly is no 2.4-million-year-old volcanic tuff at Koobi Fora, as both Brown and McDougall contend, and if Mitchell's date of 2.4 million years on the infamous Leakey I sample crystals is really correct, then the KBS Tuff controversy appears to be left—in the technical realm, that is—with an unresolved mystery.

What, then, of the sociological realm, such as the intense sense of identification with the Fitch/Miller date that developed in the Koobi Fora research group—"my team," as Leakey refers to them? How did this contribute to the affair? "Some people thought the competition between the Omo and Koobi Fora people was a bad thing,"[80] observes Leakey. "But I didn't. Nor did Glynn. You know, it gives some incentive, some force to teamwork."

Anyone who has had experience of field camps will acknowledge that none is run like Leakey's. In addition to the practical side of life in the Leakey camp, which is excellent, there is demanded a very special form of commitment. Ian Findlater, who was part of the Leakey team for more than five years, describes it this way: "Richard ran an expedition and as joint leader and main operator of the practical side he felt that he had a right to loyalty from the expedition members. Inevitably that meant agreement with him on all important factors associated with the expedition. In fact, I suspect that it is the only way such an expedition can be run. Glynn [Isaac]'s democratically run side of the expedition was always chaotic and badly organized. Richard's way has its weaknesses—if you did not agree on important issues you could either back down or leave. Most of us backed down a few times and then eventually left. . . . Despite this, my own preference would be to work for an expedition run by Richard."[81]

This brand of loyalty has as much to do with Leakey the man as with Leakey the leader of a team of scientists. In fact, it may be a brand that can at times be incompatible with the way a group of scientists must operate. It does not allow for an independence of mind and outspokenness that are essential in scientific advancement. Leakey himself describes it as "an old-fashioned kind of loyalty."[82] Andrew Hill, who worked at the museum in Nairobi for much of the 1970s, says that "It is something genuine and nonmanipulative. It is not as crude as 'You do as I say and get the results I want,' as some of his critics hold. It is a genuine loyalty, which he gives to his colleagues too."[83] In fact, many people see Leakey's bestowal of this unquestioning loyalty as part of the KBS Tuff problem, and Leakey acknowledges the possibility: "Jack and Frank were on my team and I supported them perhaps longer than I should have."[84] By being loyal to particular people in this situation, Leakey in effect was committing himself to a particular experimental result, which proved injurious.

But, as Andrew Hill notes, there was an unmistakable attraction in Fitch and Miller's particular results. Not only did they represent "numbers," which many perceive as inherently more scientific than "mere interpretation of fossils," but also, he points out, "Once you have a result like that, it is understandable that you'd want to hang on to it for as long as possible. It was very gratifying to the expedition, because it is always useful to have the oldest *Homo*, the oldest artifacts, and so on."[85] Leakey denies that dates as such are important to him; he maintains that he is more interested in why *Homo* developed a large brain than in when. Everyone in the business is aware, however, that an older date is "better" than a younger one, if for no other reason than for fundraising, an endeavor in which Leakey has consummate skill.

From Leakey's point of view, the KBS Tuff controversy was educational, if nothing else. "It taught me a lot about the scientific community, in hindsight,"[86] he now comments. "One realizes that even in the most pure of sciences, which geophysics should be, there is a potential to identify careers, status, and results—and there's a strong political element, too. I should have known this, because I had never really developed the respect that I suppose I should have done for science. But I was upset at times to realize that we may have been given a line that wasn't necessarily secure, even in their own minds."

Leakey is frank about his lack of technical training. "I am not a proper scientist, never will be." Although Leakey sometimes wears this like a badge, his friends and colleagues believe that in the KBS affair, at least, it was a liability. "I felt that Richard wasn't capable of making certain judgments, unless he had all the evidence,"[87] comments Leakey's erstwhile colleague and rival Clark Howell. "And it was very one-sided evidence that was coming through." "Brilliant as Richard is, he has not learned how to be a scientist"[88] is Garniss Curtis' assessment.

"I was very young and ill-equipped to handle what developed

into a mighty issue,"[89] Leakey concedes. "I did not have the authority, background or training to be able to evaluate what was going on. I was really out of my depth with these dating techniques, but I felt I hadn't anyone else to turn to. It would have been too humiliating to turn to Clark Howell. I think a bit of humiliation is good for one now. But not then."

POSTSCRIPT: While I was researching the history of the KBS Tuff controversy, I discovered that a few of the original crystals—the so-called Leakey I sample—collected from the KBS Tuff still existed. These crystals, remember, were the ones that were dated by Fitch and Miller at 2.61 (later revised to 2.42) million years and, according to Frank Fitch, "led us astray for so long." I discovered the existence of the crystals while interviewing Tony Hurford, who, after he left Fitch's lab in London, had taken them to the University of Berne, Switzerland, where he now runs a geological dating laboratory. Here, it seemed, was an opportunity to settle a key unsolved mystery at the center of the KBS Tuff affair, so I asked Hurford if he would run a potassium/argon dating test on the crystals. He agreed.

A date close to 2.4 million years would support Fitch and Miller's contention that the crystals had come from a tuff other than the KBS, that they and everyone else had indeed been misled by being provided with incorrectly collected material. A date close to 1.9 million years, however, would be consistent with the crystals having truly come from the KBS Tuff, the erroneous older date having been the result of unexplained problems in the original dating procedure.

Hurford wrote to me on 11 February 1987 with the first preliminary results: "The Leakey sample gave 1.87 ± 0.04 [million years]," he noted. Further work would be required to substantiate these results, he said, but "I don't expect much change—certainly not 2.4 Ma!" Further results were forthcoming at the end of February, and they confirmed the 1.87-million-year-date.

A feldspar crystal, 8 mm. in length, from the original KBS Tuff sample, which gave a date of 2.61 million years in the Cambridge Laboratory. "These crystals are what led us astray for so long," says Frank Fitch.
© Frank Fitch

John Harris measures a *Deinotherium* tooth during the 1968 expedition to Koobi Fora. With Richard Leakey and Kay Behrensmeyer, Harris for a time supported the idea that different rates of evolution might explain the differences between the animals at Koobi Fora and the nearby Omo. "I can see now that we were seeking ways of justifying that date rather than objectively trying to clarify the evidence," he says.
© Bob Campbell

Frank Fitch, who, with Jack Miller, was responsible for producing the dates of the volcanic layers at Koobi Fora. In August 1969 he told Richard Leakey that using their newest technique "would result in [the KBS Tuff] being incontrovertibly dated and with greater accuracy than any other site in Africa or elsewhere." However, controversy, not incontrovertible dating, was what followed. © *A. H. Hurford*

F. Clark Howell, co-leader of the Omo Valley expedition and "boss of the rival gang." Howell was concerned that the lists of fossil animal species at the Omo and Koobi Fora did not match up as they should—if the 2.6-million-year date was correct for the KBS Tuff. Before the Nairobi conference he discussed the problem with Frank Brown and concluded that "there's got to be something wrong with the KBS date." © H. B. Wesselman

Bernard Ngeneo, the man who found 1470, also discovered the pelvis that at one time was thought to be at least 3 million years old. Richard Leakey and Glynn Isaac published a paper saying that the pelvis was "dramatic confirmation" of the existence of ancient *Homo* at Koobi Fora, even if the cherished 2.61-million-year date for the KBS Tuff proved to be wrong. © Bob Campbell

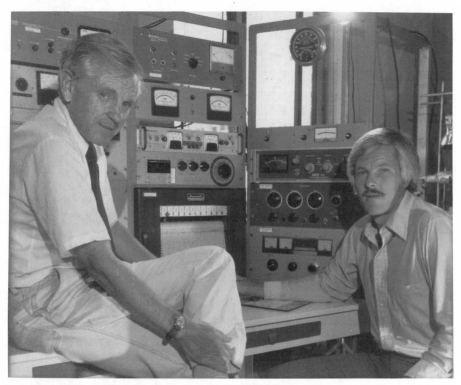

Garniss Curtis (left) and Robert Drake, who produced the initially unacceptable 1.8-million-year date for the KBS Tuff. Curtis wrote to Leakey in August 1975, "I'm married to a clinical psychologist who constantly points out to me how unobjective scientists are in general and unobjective I am in particular; however, I believe that I have *tried* to be objective in our paper concerning the age of the KBS Tuff." © *Joachim Hampel*

The Koobi Fora "summit" of August 1976. Ian Findlater, center, explains his interpretation of the Koobi Fora geology to (left to right) Jack Harris, Kay Behrensmeyer, John Harris, Glynn Isaac (obscured by Harris), Richard Leakey, and Meave Leakey. "Ian was unhappy about having his conclusions questioned," says John Harris. "And we were very unhappy at having to fit what seemed like a logical evolutionary sequence into a scheme that made nonsense of it." © *T. D. White*

Ian McDougall, an Australian geochronologist whose radiometric data did much to clarify the time scale at Koobi Fora, wrote the following of the problems with the KBS Tuff date: "The large spread of results reported by Fitch and Miller must result from experimental difficulties or error estimations that do not properly reflect the uncertainties in the actual measurements." © *Ian McDougall*

Richard and Meave Leakey visit the first—and infamous—KBS sample site in January 1985, at the request of the author. Responding to Jack Miller's suggestion that his and Fitch's erroneous dates for the KBS Tuff were the result of material's having been mistakenly collected from other tuffs, Richard Leakey says: "It is a most extraordinary explanation. I am as sure of where that first sample site is as I am about where my house is." © *R. Lewin*

Basil Cooke (left) examines pig fossils with Richard Leakey at Koobi Fora in 1973. It was Cooke's data on fossil pigs of the Omo, Koobi Fora, and Olduvai that initially threw doubt on the validity of the 2.6-million-year date for the KBS Tuff. After expending several years of effort on the project, he was effectively eased out and replaced by John Harris and Tim White. "My only real consolation was that their study, commissioned to refute my dating contentions, in fact only served to confirm them," he now says.
© *H. B. Wesselman*

Frank Brown, like several other geochronologists, was very much puzzled when Fitch and Miller recomputed their 2.6-million-year date for the KBS Tuff to be 2.4 million years, which coincided with the just-published date produced by the fission-track method. "I didn't receive that well at all," Brown now says. © *H. B. Wesselman*

Glynn Isaac, left, was co-director with Richard Leakey, right, of the Koobi Fora Research Programme, and was Leakey's staunchest ally. Nevertheless, in June 1975 Isacc realized there was a problem with the 2.6-million-year date for the KBS Tuff. "Glynn has totally abandoned ship," Leakey wrote to a colleague when Isaac told him he thought Fitch and Miller might be wrong. Yet largely because of his strong sense of loyalty to Leakey, Isaac continued publicly to support him in his championing of the Fitch/Miller date.
© H. B. Wesselman

Donald Johanson announces the new hominid species *Australopithecus afarensis* in Stockholm, May 1978. "I finished my talk, looked up at the audience, and everyone just sat there," recalls Johanson. "I was shocked. It was as if they had all decided they would ignore it." © *John Reader*

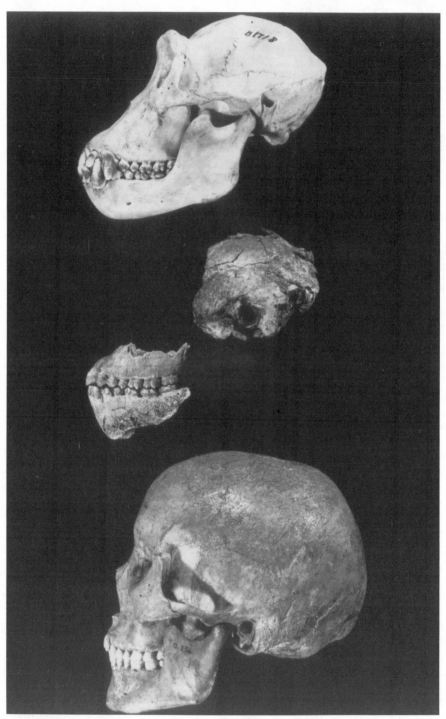

Australopithecus afarensis, although it is a hominid, is much more like a chimpanzee (top) than a modern human (below). Many anthropologists consider that Lucy and her fellows must be very close to the point at which the human and ape lines split. © *Cleveland Museum of Natural History*

Mary Leakey felt strongly about the naming of the Hadar and Laetoli fossils. "I do not think *Australopithecus* is correct," she told Johanson. "It is a lousy term, based on a juvenile. . . . Nor is it a direct ancestor of *Homo,* as all of us people agree." *© Delta Willis*

Tim White inspects his reconstruction of the skull of *Australopithecus afarensis,* which was based on 107 small fossil fragments. *D. C. Johanson © Institute of Human Origins*

The Ethiopian fossils. In December 1975, shortly after the "First Family" fossils had been uncovered at the Hadar site, Johanson took the fossils to Nairobi. Richard and Mary Leakey said that some of the specimens appeared to be early *Homo* while others were early *Australopithecus.* "I was inclined to agree with them," Johanson later said. Tim White, however, was not convinced. "We should at least consider the possibility that we are dealing with one thing here rather than several taxa," he said. Here, White explains his ideas to a skeptical Leakey, Johanson watches the exchange, and Bernard Wood ponders. *David Brill © National Geographic*

The Hadar site, Ethiopia. Donald Johanson's tented camp, which was organized jointly with French anthropologists Maurice Taieb and Yves Coppens, was situated on the banks of the fast-flowing Awash River. This view is looking west, over the dust-dry sediments of an ancient lake. *David Brill © National Geographic*

Mary Leakey (second from right), stands with Donald Johanson on her right, as she prepares to accept the Golden Linnaean Medal, presented by King Gustav of Sweden at the May 1978 Nobel Symposium. © *John Reader*

The Kirtlandia paper which, as a result of the displayed telegram, was withdrawn at the last minute so that Mary Leakey's name could be taken off the title page. "They put my name on the paper without my permission," contends Leakey. The lower jaw is Laetoli hominid 4 (LH4), the type specimen of *Australopithecus afarensis.* © *T. D. White*

Tim White excavates hominid footprints at Laetoli in the summer of 1978. Shortly after this picture was taken, White left Laetoli following a vigorous disagreement with Mary Leakey over the naming of *afarensis* and the inclusion of her name on a scientific paper. White never returned to Laetoli while Leakey was in charge.
© *Peter Jones*

Footprints in the sands of time. This trail of hominid footprints at Laetoli, Tanzania, is dramatic demonstration that our ancestors walked upright 3.75 million years ago.
© *John Reader*

Ernst Mayr, one of this century's greatest evolutionary biologists, is critical of the choice of the technical name for the Lucy species, coupled as it is with a "type specimen" from Laetoli, in Tanzania. "You cannot" he now says, "have things that are in totally different localities and different times, then pick the name from one place [the Afar] and the type specimen from the other [Laetoli]. © *Harvard University*

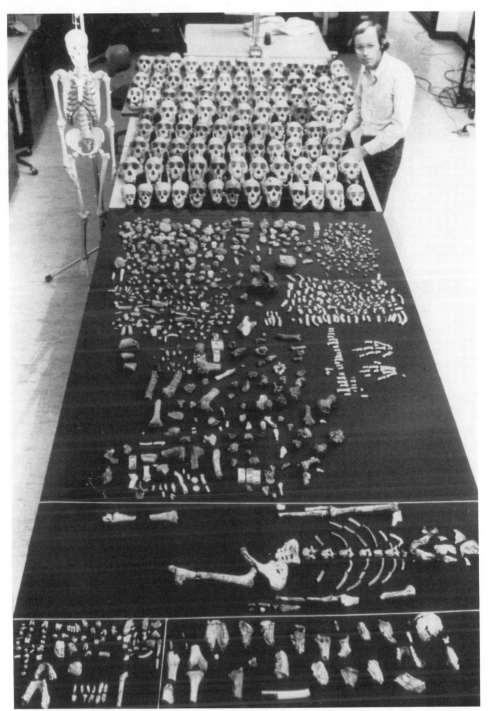

Fossils galore. Laid out in front of the ranks of chimpanzee skulls are, first, the remains of the "First Family" from the Hadar—a cache of several hundred bone pieces from at least a dozen individuals; in front of this remarkable collection is Lucy; in front of Lucy, at right, are more *Australopithecus afarensis* fossils, including the famous knee joint; and at bottom left are the Laetoli specimens, fragments of perhaps another 13 individuals. *D. C. Johanson* © *Institute of Human Origins*

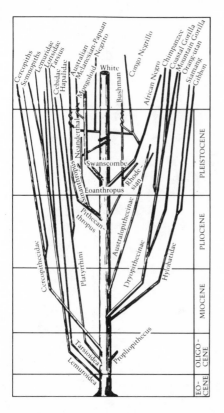

Racism in anthropology. Many of the human family trees produced during the early decades of this century show British and American anthropologists' predilection for seeing white, Nordic races as the pinnacle of human evolution. Four examples are given here: 1. Earnest Hooton's, in his book *Up from the Ape*, first published in 1931. 2. Grafton Elliot Smith's, in his book *Human History*, first published in 1930. 3. Henry Fairfield Osborn's, as produced for the American Museum in 1923. And 4. William King Gregory's, in his book *Our Face from Fish to Man*, published in 1929.

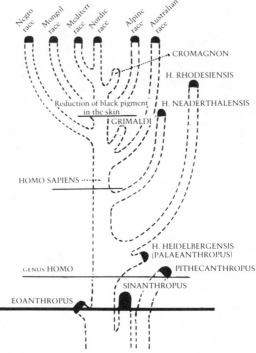

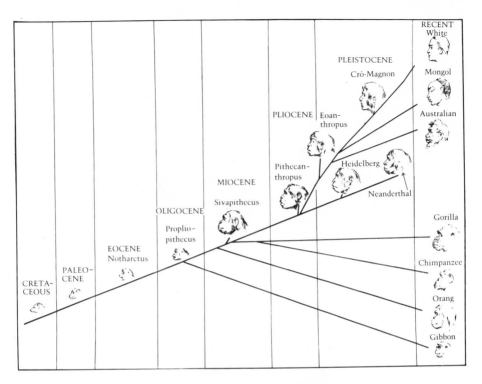

RECENT
White

PLEISTOCENE
Crô-Magnon

Mongol

Australian

PLIOCENE | Eoan-
thropus

Pithecan-
thropus

Heidelberg

MIOCENE

Sivapithecus

Neanderthal

OLIGOCENE

Proplio-
pithecus

Gorilla

Chimpanzee

EOCENE
Notharctus

Orang

PALEO-
CENE

Gibbon

CRETA-
CEOUS

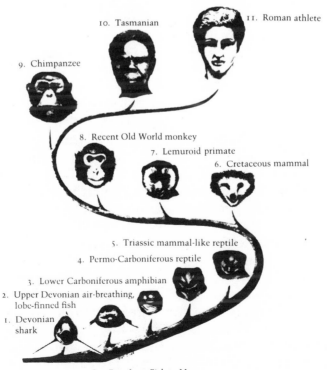

11. Roman athlete

10. Tasmanian

9. Chimpanzee

8. Recent Old World monkey

7. Lemuroid primate

6. Cretaceous mammal

5. Triassic mammal-like reptile

4. Permo-Carboniferous reptile

3. Lower Carboniferous amphibian

2. Upper Devonian air-breathing,
lobe-finned fish

1. Devonian
shark

Our Face from Fish to Man

CHAPTER 11

LUCY: THE NAMING

"It was a tremendous thrill for me, a great reward," recalls Don Johanson. "There I was, the son of a Swedish immigrant who was a cleaning lady. And I had been invited to speak at the Royal Swedish Academy of Sciences, to participate in a Nobel Symposium that was to be attended by the King and Queen of Sweden."[1] The gathering, which was held in May 1978, was part of that country's commemoration of the two-hundredth anniversary of the death of Carolus Linnaeus, the great Swedish scientist who in 1758 established the system by which relationships between all living organisms can be systematically organized and labeled.

For Johanson, whose parents had emigrated to the United States from Sweden, the emotion attached to returning to his "homeland" under such circumstances can readily be imagined. "My relatives were very proud," he recalls.

The honor that Johanson experienced in taking part in this important symposium was redoubled by the message he was there to deliver. He was to utter for the first time in public the words *Australopithecus afarensis*, which was the name of the first major new hominid species to be designated in fourteen years. Johanson, together with his colleagues Tim White and Yves Coppens, was the author of that new name. Moreover, Johanson considered that *Australopithecus afarensis*, which had lived between 3 and 4 million years ago and includes the famous Lucy skeleton, was the ancestor of all later hominids, the rootstock of us all.

During the fourteen years since Louis Leakey announced *Homo habilis*, in 1964, there had been a series of remarkable discoveries of fossil hominids—in Tanzania, in Kenya, and in Ethiopia. It had been an unprecedented era of finds, and those in the paleoanthropological profession were in a state of great excitement and some puzzlement over what all these fossils might imply for their hypotheses. Johanson therefore confidently expected his message to be received with enthusiasm, if not with immediate acceptance.

"I finished my talk, looked up at the audience, and everyone just sat there," recalls Johanson. Silence. "After a while, someone said, 'Well, if there are no questions, we'll break for coffee.' Everybody left the room, except David Pilbeam, who congratulated me; Yves Coppens, who said, 'Well done'; and Phillip Tobias, who said, 'My God, I was going to name two subspecies and you've scooped me.' There was no discussion of the new species. I was shocked. It was as if they had all decided they would ignore it."[2] Richard Leakey was there. Mary Leakey was there. They said nothing—nothing in the manner of scientific discussion, that is. Mary Leakey was angry.

She strode up to Richard during the coffee break and fumed. "Did you hear that?" she demanded rhetorically. "That fellow talked about my fossils. He talked about my site. How am I going to give my paper now? It's all been said."[3] Richard could only suggest that she should say it better. "Mary doesn't like public speaking, and it was going to look as if she was a fool, repeating the same material," he recalls.

There are unwritten rules of etiquette governing how one scientist might discuss in public the data of another scientist. And for paleoanthropologists, fossils are data. If certain fossils have not yet been published in a scientific journal, no one may discuss them in public without the clear permission of the discoverer or the person analyzing them. And it is a rare occasion when this might happen. Once published, and therefore in the public domain, however, data may legitimately be addressed by anyone—given the proper attributions, of course.

In the case of the Nobel Symposium, the information relating to Mary Leakey's site and her fossils had already been published as a scientific article in *Nature* two years earlier, as Johanson points out. "None of the things I spoke about were unpublished or unavailable to any of the audience," Johanson now says.[4] In spite of assertions to the contrary, therefore, Johanson appears not to have committed a scientific impropriety in discussing Mary Leakey's data at that meeting. However, it could be argued that common courtesy had been overstepped, as the author of the data was due to speak about them at the same meeting.

But Johanson had a compelling reason for describing the Leakey fossils in his talk. His new species, *Australopithecus afarensis*, was based not only on his fine collection of fossils from Ethiopia but also on Mary Leakey's much smaller but nevertheless important group of specimens from Laetoli in Tanzania. And one of Mary Leakey's fossils, part of a lower jaw, labeled LH-4 (for Laetoli hominid 4), was to take pride of place in the naming of the new species. Whenever a new species is erected, the author must designate what is called a "type specimen," a kind of flagship for the species, a fossil against which all others must be compared if possible. Johanson named LH-4 as the type specimen of *Australopithecus afarensis*: as those in the nomenclature trade know it, LH-4 was the "name-bearer" of the new species. Therefore he had to talk about Mary Leakey's Laetoli site and her fossils in his presentation.

All this might sound somewhat arcane, and it is true that the conventions for the proper naming of fossils are about as esoteric as one can get without becoming mired in something like chancery law. But the naming of *afarensis* was to be at the center of what quickly developed into a mighty and very public row.

On the face of it, the row was about the interpretation of the Laetoli and Hadar fossils, by Johanson and White on one hand and Mary and Richard Leakey on the other. But other issues fold into the mix: specifically how the two groups of fossils—one set from Tanzania, the other from Ethiopia—came to be joined in what developed into an uncomfortable alliance.

Mary Leakey had collaborated freely and congenially with Johanson over their fossils for some years, but the Nobel Symposium marked a watershed in their relationship. Until that point—May 1978—Mary Leakey apparently supported Johanson's position, with some qualifications. But after the symposium she abruptly and dramatically began to dissociate herself from Johanson—an action that undoubtedly influenced the reception of *afarensis* by the paleoanthropological community. Why and how this occurred is therefore of more than a little interest.

Also of importance was the fact that by setting up *Australopithecus afarensis* as the ancestor of all later hominids, Johanson specifically "demoted" another species, *Australopithecus africanus*, which in many people's eyes had previously occupied that position. The principal champion of *africanus* was Phillip Tobias, the man who now occupies Raymond Dart's professorial chair in Johannesburg and the man whom Johanson "scooped" at the Nobel Symposium. Tobias was not pleased, and ever since has done all he can to oust the usurper. More than once he has used international gatherings formally to propose the suppression of Johanson's new species and the reinstatement of *africanus*, so far without success.

How Lucy got her formal scientific name, and the paleoanthropological community's response to it, is therefore a complex story. In paleoanthropology, the answer to that old question "What's in a name?" is "Everything." By giving a fossil a new name you do at least two very important things. First, you make a statement about how the fossil might fit into the current picture of the human family tree. In this case, the scientific naming of Lucy completely overturned previous ideas about human origins—with the inevitable result that some sections of the paleoanthropological community were distinctly upset. No scientist likes to see his pet theory swept aside, and this is especially so in paleoanthropology, where individual researchers tend to be more intimately involved with and proprietary about their theories than in other sciences. And second, by giving new fossils a new name, the author of the name is forever linked with those fossils, to the exclusion of all other scientists. In this case, Mary Leakey saw "her" fossils being appropriated by Johanson and merged into anonymity with the Hadar fossils. And she did not like it.

The story of how Lucy got her name is therefore more than an account of a scientific christening. It is a confection of professional and personal responses to an intellectual upheaval in the field. It is a story that, with varying degrees of clarity, reveals the swell of underlying preconceptions as they bolster at least three different intellectual positions. The process by which the naming took place is described in this chapter. The aftermath of the naming, which continues to this day, is recounted in the following chapter.

When, in the late summer of 1973, Johanson set out for Ethiopia as a co-leader of the Joint International Afar Research Expedition, he was an unknown researcher from Cleveland, Ohio, with a yet-to-be-completed doctorate. Looking like a "nail-polish salesman, wearing Yves Saint Laurent pants and Gucci sneakers," as his friend and colleague Tim White once noted, Johanson doesn't immediately give the impression of one who would adapt to the rough, tough life of fossil-hunting fieldwork. Especially not in the grueling badlands of Ethiopia's Afar region. But within four years Johanson not only had displayed a talent for field organization that approaches that of his friend and friendly rival Richard Leakey, but also could boast, with his French co-leaders Yves Coppens and Maurice Taieb, one of the most impressive and important early-hominid collections ever recovered in Africa. This includes Lucy, an unprecedentedly complete (40 percent) skeleton of a very early hominid; a cache of some 350 fossil fragments that represent at least 13 individuals and has come to be known as the First Family; several other sets of jaws and teeth, and a beautiful little knee joint.

As Duke University anthropologist Matt Cartmill notes, "any one of these finds would be enough to catapult the finder into international prominence and turn competitors green with envy."[5] What collectively they did for Johanson was transform him "from a brash young Ph.D. with a nervous grin and a promising Pliocene site into a paleoanthropological superstar with a trunkful of the most dazzling fossil-hominid jewels ever to come out of East African hominid mines. . . . Johanson's competitors and colleagues, myself included, have been Day-Glo chartreuse for the past five years," adds Cartmill.

It also transpired that Johanson had a talent not only for finding fossils but also for making the most of the public interest that the fossils inevitably generated. As a result, he has sometimes been called the Carl Sagan of anthropology. And like Carl Sagan, Johanson has found that success in the public eye has provoked some professionals to charge him with self-promotion. "A lot of that can be traced to jealousy,"[6] says Tim White. "A lot of them would like to be Don Johanson and the rest would like to be Richard Leakey," he adds, which is a measure of Johanson's meteoric rise.

While this public acclaim still lay ahead, Johanson and his colleagues began their preliminary assessment of what the fossils meant. They knew that the Hadar hominids were quite primitive in appearance and were a little over 3 million years old, judging from potassium/argon dating tests done by Jim Aronson of Case Western Reserve University. This made them at least a million years older than Mary Leakey's fossils at Olduvai Gorge. How they matched temporally Richard Leakey's fossils from Koobi Fora was uncertain, because at the time the KBS Tuff controversy was still unresolved. Nevertheless, the evolutionary pictures at Olduvai and Koobi Fora were similar: at least two species of hominid—one *Homo* and one *Australopithecus*—had coexisted at these two sites. The question was, how did the new fossils from Hadar fit in?

Perfectly, it seemed. In his first full publication on the Hadar fossils, which appeared in *Nature* on 25 March 1976, Johanson wrote with Taieb that "On the basis of the present hominid collection from Hadar it is tentatively suggested that some specimens show affinities with *A. robustus,* some with *A. africanus,* and others with fossils previously referred to *Homo.*"[7] In other words, just as at Olduvai and Koobi Fora, the Hadar site had *Australopithecus* and *Homo* living side by side. The idea that both species of *Australopithecus*—the gracile and the robust—should be present at the same site in East Africa was somewhat novel, but it was an idea soon to be embraced for Koobi Fora.

Some years after that *Nature* article had entered the scientific literature, Johanson commented ruefully, "I would withdraw that paper today if I could. It stands as an object lesson to me not to be too hasty in the future."[8] At the time, however, everything seemed to fit nicely.

Indeed, one of *Nature*'s anonymous paleoanthropology correspondents would soon make the following observation: "The Lake Turkana and Hadar discoveries have provided virtually unquestionable evidence that at least two forms of hominid had coexisted. . . . The second implication of the new fossil material is that the genus *Homo* may be far older than previously thought."[9] This line of thinking—that the origins of *Homo* are to be found far back in the fossil record—has come to be closely associated with the Leakey name, first with Louis and then with Richard.

Johanson's assessment was not carried out in isolation, of course. He had frequent contact with the Leakeys throughout those crucial years of 1973 to 1977. Mary and Richard once flew to the Hadar site, in November 1974, and were very much impressed with what they saw: clear evidence of *Homo,* they opined, in several jaw fragments. And Johanson, together with each season's haul of hominid fossils, routinely traveled back to the United States via Nairobi. One reason was purely practical and precautionary: the casting technicians at the Kenya National Museum in

Nairobi were able to make a set of excellent replicas of the Hadar fossils, which were kept in Nairobi just in case Johanson's plane went down on the journey home. But he also enjoyed the opportunity to show Richard the new material and discuss it with him.

Johanson's 1975 pass through Nairobi was particularly rewarding, because that season had seen the discovery of the so-called First Family. Never before had such a collection of early hominid fossils been discovered: more than 300 fossil fragments—parts of jaws, crania, hands, feet and limbs—and all apparently deposited within a short period of time. In fact, Taieb even considered that the thirteen individuals in the assemblage might have died at the same time, perhaps the victims of a flash flood. Johanson embraced the suggestion, saying that "it is the earliest direct evidence of what we know must have been the case, that these hominids were social creatures, living in groups."[10] It was certainly a seductive idea, that these individuals were perhaps a family and met a collective, tragic end. That would be extremely important scientifically, because for the first time paleoanthropologists would be able to get an idea of anatomical variation within a true population of related individuals. "And as all ages were represented in the group, from infants to adults, we can get an idea of how the anatomy changed as individuals grow up, how it changes ontogenetically," Johanson observed shortly after the discovery. But it now seems much more likely that instead, the bones accumulated over a period of many years and are not the remnants of some paleo-drama.

Nevertheless, as they lay on the specimen table in the Nairobi Museum that 30 December 1975, they were the subject of a modern drama. "When I spread out the haul of new bones, they were an instant sensation,"[11] recalls Johanson. "Nothing that combined their extreme antiquity, their remarkable quality and their profusion had ever been encountered before." Although the fossils had yet to be properly cleaned and prepared for analysis, it was very clear that most of these individuals had been much bigger than Lucy and her kind—up to twice as big in some cases. And it was clear that they reflected a curious mixture of primitive and advanced features.

Sitting, standing, and leaning over the table, talking with a quiet but obvious excitement, were Johanson, Richard and Mary Leakey, Phillip Tobias, and Bernard Wood. Fossils were lifted to the light, turned to favorable viewing angles, passed carefully from hand to hand. All declared that the fossils represented a fine collection, very probably among the earliest examples of the genus *Homo* yet discovered. Lucy, though, was surely something different, perhaps something new. Later there developed the idea that Lucy was a remnant of a much earlier, primitive form, "a terminal occurrence of an ancient type,"[12] as Johanson put it. But at that initial display of First Family fossils in Nairobi, the most important and powerful message that Richard and Mary Leakey perceived was that *Homo* and some form of *Australopithecus* had coexisted more than 3 million years ago. "I was inclined to agree with them," Johanson noted later.[13]

Another scientist was present at that gathering, but unlike the others, he was rather quiet and did not join in the general discourse. This was Tim White, who at the time was a graduate student working on a dissertation on early-hominid jaw structure and function and had just completed a second field season at Koobi Fora. Those who know White know that he is rarely quiet. Johanson hadn't met White before and mistook his silence for timidity. "I wasn't timid, for God's sake," he later told Johanson. "I was just being sensibly cautious. Here you were, the smooth young hotshot, shooting your mouth off about all your great fossils. I'd never met you before. I didn't know if you could tell a hippo rib from a rhino tail. I was just waiting for you to fall flat on your face, say something really dumb."[14]

And so began a most productive and important professional relationship, one that was to have a tremendous impact on the science of paleoanthropology and on the community itself.

White was doing his thesis from the University of Michigan, under Milford Wolpoff as supervisor. At the time, Michigan was the center of a particular view of human evolution, known as the single-species hypothesis. The hypothesis was the conceptual product of Loring Brace, a paleoanthropologist remarkable for his unique historical perspective on the science. He also has a sharp intellect and a sharp and sarcastic wit. Wolpoff was a devotee of Brace's hypothesis, which proposed that the anatomical differences seen between the various hominid fossils recovered from South and East Africa were the result of variation within a single species, not the distinguishing marks between several species. In other words, said Brace, there was just one hominid species in Africa at any one time, which gradually became more and more advanced.

White found the hypothesis difficult to accept, especially after he had visited Africa and seen the original specimens for himself. As a result, he failed one set of written exams on fossil man, which was part of his doctoral course. "I demanded to do the exams again, this time in an oral format,"[15] says White. "We all sat around a big table, and when we disagreed, I said, 'Let's go to the cabinets and get the casts; let's go to the literature so we can check the numbers.' I passed this time." Although he balked at the pure form of Brace's single-species hypothesis, White may well have absorbed something of the Michigan school's ethos. For it was White who would eventually argue that the Hadar fossils represented just one species, not two or three as the Leakeys and Johanson were proposing. But this still lay ahead.

When he first went to Kenya, in 1974, to work on his dissertation, White quickly impressed the Leakeys and their colleagues with his congeniality and his talent for anatomical description of fossils. It so happened that during the 1974 and 1975 field seasons Mary Leakey and her colleagues had recovered a series of fossil hominids—mainly teeth and jaws—from Laetoli, a remarkable site 40 kilometers south of the famous Olduvai Gorge. During 1975, Garniss Curtis and his Berkeley colleague Robert Drake had established the age of the fossils as between 3.59 and 3.77 million years, using potassium/argon dating techniques. The Laetoli fossils were therefore the oldest undisputed hominids ever discovered, and Mary Leakey wanted to have someone good to do the descriptions for publication.

"When Mary asked me who might do the descriptions," recalls Richard Leakey's close colleague Alan Walker, "I said, 'Why don't you give them to Tim? He's very good. He's working on mandibles and the Laetoli fossils are mainly jaws and teeth'."[16] Richard Leakey also encouraged Mary to ask White, which she did, on 18 November 1975. "It was a good choice," says Walker, "a very good choice." Mary Leakey's view of White at the time was that he was "a good scientist and a hard worker, if sometimes a little naive."[17]

So it was that just one month before Johanson brought his First Family fossils through Nairobi, Tim White began a detailed study of Mary Leakey's Laetoli fossils. The timing proved to be important. "When I saw Don's fossils I could see that they were the same as Laetoli—whatever it was,"[18] recalls White. "I hadn't seen the original Lucy fossils, just a cast. But from what I had seen, there was nothing to exclude her from being grouped with the First Family fossils and the Laetoli fossils. Don said that Lucy was too small, too different to be put in the same group as the others. But it was my approach to say they are the same, to expect there to be differences between males and females. So I said to Don, 'We should at least consider the possibility that we are dealing with one thing here rather than several taxa.' His response was 'No way.' "

White left it at that, and continued working on the description of the Laetoli fossils, which was completed by the end of January 1976. The paper containing White's work, together with the geology and paleontology of the Laetoli site, was published in *Nature* six months later, on 5 August. The interpretation of the hominid fossils was quite clear. "Preliminary assessment indicates strong resemblance between the Laetoli hominids and later radiometrically dated specimens assigned to the genus *Homo* in East Africa."[19] In other words, the Laetoli jaws and teeth look something like the *Homo habilis* fossils from Olduvai and 1470 and its fellows at Koobi Fora. There were differences, however, which the paper addressed as follows: "It should come as no surprise that the earliest members of the genus *Homo* display an increasing frequency of features generally interpreted as 'primitive' or '[ape]-like', which indicate derivation from as yet largely hypothetical ancestors."

With the benefit of hindsight, White explains that he wrote these words at a time when there was a tradition that hominids at the East African sites were one of two types: either the big, robust *Australopithecus* or the smaller *Homo*. "It was a way of thinking at the time,"[20] he recalls. "So when these little jaws came out of Laetoli it was 'obvious' they must be *Homo* because they certainly weren't the large *Australopithecus*." How much White was part of the tradition of the time is difficult to discern. He says, "I knew what words would be acceptable and what wouldn't," but "I was unconscious of the path I was following."[21]

So at the beginning of 1976 the pattern of human evolution in East Africa—on the basis of the Olduvai, Koobi Fora, and Hadar fossils—seemed to be relatively clear and internally consistent. To wit, both the *Homo* and *Australopithecus* lineages had arisen by at least 3 million years ago, the evolutionary descendants of a yet-to-be-discovered common ancestor. The next two years were eventful for all parties to this scenario: for Richard Leakey, the KBS Tuff controversy moved toward resolution; for Mary Leakey, there was the discovery of perhaps the most remarkable of all signatures in the fossil record, hominid footprints dating back more than 3.6 million years; and for Johanson, a final field season at Hadar yielded more fragments of the First Family, and the world's oldest stone tools. Most significant of all, however, was White's success in persuading Johanson that his Hadar fossils belonged to a single species, which was ancestral to all later hominids.

The footprints at Laetoli are the product of a remarkable confluence of circumstances. First of all, the ash of a nearby volcano, Sadiman, contains a high concentration of carbonatite, which sets like concrete when wetted and then dried. A little more than 3.6 million years ago Sadiman spewed a layer of its unusual ash over the southern Serengeti, and soon thereafter a brief, light shower fell. Some of the raindrops punched droplet craters in the ash, which are preserved to this day. While the ash layer was still damp, a veritable menagerie of twenty different species of animal walked, ran and slithered over it: these included hares, baboons, several antelopes, an elephant relative, two types of giraffe, a saber-toothed cat, hyenas, a curious claw-footed ungulate, many birds—and hominids. The imprinted layer was soon covered by more ash and windblown silt, there to remain preserved and unseen until 15 September 1976.

Andrew Hill, a British paleoanthropologist then based in Kenya and now at Yale University, discovered the first (nonhominid) prints on that day, when his eyes came to rest a few inches from the recently exposed ash layer. Although his propitious posture was the result of a rapid evasive maneuver designed to avoid impact by a large lump of elephant dung playfully hurled at him by biologist David Western, rather than an instance of close paleontological prospecting, it was nonetheless effective.

If the ash layer had contained only animal prints, it would have been extremely valuable scientifically in capturing a glimpse of a true paleo-ecological community in the way that fossils can only hint at. But that it also bore trails of hominid prints catapulted it into a quite different dimension. Mary Leakey describes the hominid trails as "perhaps the most remarkable finds I have made in my whole career."[22]

Although there is still some discussion about their precise interpretation, the individual prints in the hominid trails are strikingly humanlike. Sight of them as they wend their way across that 3.6-million-year-old Laetoli landscape is an inescapably moving, even eerie, experience. Not only are the prints "remarkably similar to those of modern man,"[23] says Mary Leakey, but they "could only have been left by an ancestor of modern man." Therefore, for Mary Leakey the discovery of the footprints served to support the conclusion that the teeth and jaws described by White were indeed from members of the genus *Homo*. "The form of the prints fully confirms this,"[24] she says.

This line of argument rests on the assumption that only species of *Homo* would have feet and a gait like ours and that the footprints of *Australopithecus* would be somehow identifiably different: more primitive, perhaps. It is an assumption of the sort that has often been made in paleoanthropology, but it appears to be based as much on special pleading—a kind of homocentrism—as on hard evidence.

Meanwhile, 1,500 kilometers to the north, Johanson and his colleagues were also making discoveries in Ethiopia that spoke of *Homo*. During the 1976 season Hélène Roche, a French archeologist, found a few crude stone choppers and flakes lying in a gully near the main camp. They were very much like the earliest tools from Olduvai Gorge, which are known as Oldowan, but if anything, displayed somewhat better workmanship. In order to get a date for the Hadar tools, it was necessary to recover some by excavation from undisturbed sediments: surface finds of either fossils or stone tools cannot be used reliably for dating purposes. Because Roche had to return to France, Johanson asked Jack Harris to do the job. Harris, who has worked extensively at Koobi Fora, soon uncovered tools from sediments dated at an astonishing 2.5 million years old, thus making them the oldest in the world.

"They were a stunning surprise," notes Johanson. "And they tended to strengthen my published opinion that the large hominids at Hadar were *Homo*."[25] Again, this rests on the assumption that only *Homo* was mentally equipped to fashion stone artifacts, which may be another example of homocentrism.

The discovery of the stone tools was one of the last paleoanthropological finds at Hadar, because after the 1976/77 field season, political turmoil in Ethiopia prevented Johanson and his colleagues from returning. Still, the site has yielded a wealth of material that needed to be analyzed: the skulls and teeth were to be studied under Johanson's supervision in the United States, while the limb bones would be studied by Yves Coppens and his students in Paris.

For Johanson, the collaboration with Tim White would be crucial. After their first encounter and brief discussions in Nairobi in December 1975, their next meeting was in September 1976, at an international scientific meeting in Nice, France—the one to which White had gone in high dudgeon after his blowup with Richard Leakey over the Koobi Fora pig paper. Again discussions between Johanson and White were brief, but to the point. White was now even more firmly convinced that the Hadar and Laetoli fossils represented the same, single hominid species. Again Johanson demurred: "But we have two kinds from Hadar," he insisted. "Little Lucy and the big ones." White's reply was simple: "Maybe you don't. We'll have to study that."[26]

And study it they did, in fits and starts, but always intensely, through much of 1977. White would make the trip from Berkeley to the Cleveland Museum of Natural History. The two men would share their thoughts of the previous meeting, work hard on some new aspect of the analysis, and then separate, each wondering what impact his arguments would have on the other in the quiet aftermath of the most recent intense session. The analysis crescendoed in December, with White finally persuading his reluctant colleague that the anatomical differences they saw within the Hadar and Laetoli collections were the result of variation within a single species, not the distinguishing marks between species. Big equals male, small equals female—or marked sexual dimorphism, as it is termed in the trade: this, crudely, is what the assessment concluded.

Having agreed upon this, Johanson and White then had three very practical tasks before them: first, to find a name for the species; second, to select a type specimen for the species; and third, to arrange proper announcement of the species. On each of these three issues they would eventually find themselves at odds with Mary Leakey.

Johanson, White, and Mary Leakey had frequently corresponded with each other, often about scientific issues and practical matters of field seasons. For instance, in June 1977, while still in the midst of his discussions with Johanson, White wrote to Leakey saying, "All I have are first impressions, but Lucy and the couple of other mandibles like her look to me like simply smaller versions of what Don has called *Homo*. . . . I'm finding it increasingly difficult to see any evidence that strongly supports the presence of two lineages (or more for that matter) at Hadar."[27]

By mid November, when Johanson was on the point of finally accepting White's arguments, he wrote to Mary Leakey that he was now virtually sure that the Hadar and Laetoli fossils were one species, including Lucy. He even suggested that they should therefore consider establishing a new species. "I am sure our colleagues may be somewhat taken aback by all of this, particularly Phillip [Tobias], but I have the feeling at the moment that the specimens offer our earliest evidence for the genus *Homo* and must be substantially more primitive than *Homo habilis*."[28] At this point, although Johanson was convinced that he was probably dealing with a single species, he still thought it might be *Homo*, not *Australopithecus*.

Mary Leakey's response was cautious. "I'm hesitant about any christening ceremony at this stage," she wrote to Johanson on 27 November. "We should have better cranial material first." Leakey wrote to White on the same day, with the same message. "Hope you agree—and will find it next year!" It had already been arranged that White would join the 1978 field season at Laetoli. Perhaps White would be lucky and would find a skull at Laetoli, to add to the jaws and teeth. This would certainly make Mary Leakey more comfortable with going public with a new species.

At this point—early December—White joined Johanson in Cleveland for what was to be their final analytical session. If the previous meetings had been intense, this was to be excruciating. "For 14 days we have been working nearly nonstop, sometimes until 4:30 in the morning,"[29] Johanson wrote to Mary Leakey when it was all over. He explained that they too had been hesitant about naming a new species, but that the evidence was now so overwhelming that there was simply no way of escaping it.

Johanson told Leakey that the species must be called *Australopithecus*, and explained point by point why they felt forced to that conclusion.

"Why not *Homo*?" he began rhetorically. "The Laetoli/Hadar material does not show the hallmark of the genus *Homo* as indicated by Mayr, Leakey, Tobias, Napier and others—the brain is still small and has not yet commenced enlargement. What about a new genus? This would imply that the Laetoli/Hadar hominids were significantly different from later hominids in their adaptations." Johanson also said that if they did not choose *Australopithecus* they would be compelled, for certain historical reasons, to call it *Praeanthropus*, which, he said, would be unnecessarily confusing.

"Why push for a new taxon now?" he continued. "First, and most important, we feel it is justified. . . . Second, it appears that a number of our colleagues (used in the broadest sense), based on the published material . . . are already beginning to consider dubbing a new species." Remember, once fossil data are in the public domain, anyone is at liberty to give it a new name. And provided it is done correctly, this is the name that would stick, leaving the discoverers without any of the glory.

"We feel in a nutshell, that the Laetoli/Hadar material is the oldest demonstrable hominid fossils and they are ancestral to *Homo* (brain enlargement) and *Australopithecus robustus* (dietary specialization). There is an opportunity for us to get into press soon and we would like your agreement to go ahead. The article would simply be the naming of the new taxon showing its relationships and its distinctiveness, with you and Yves [Coppens] as co-authors. I know that Yves has the idea of a new taxon. In fact, he tried to name it (twice) in the article for the Pan-African Congress [in 1977]. We have written to you first, because, frankly, we are worried that Yves may try to name a new species in France—I suspect you are fully aware of what we mean."

Coppens acknowledges that he was keen to give a new name to the Hadar fossils. "I was convinced that there was a new species among the Hadar material, from very early on,"[30] he said recently. "I was in Cleveland with Don in 1976 and suggested it then. I suggested it in September 1977, at the Pan-African Congress in Nairobi. But Don didn't want to go too fast. Yes, he was very afraid that I might name the new species."

In suggesting the name *Australopithecus afarensis* to Mary Leakey, Johanson and White knew they might face some resistance. The reason was that, like Louis Leakey, Mary has long been opposed to the idea that *Australopithecus* might be ancestral to the human line. "You've got to have *Homo* coming from somewhere, but I don't see why it should come from the southern apes [*Australopithecus africanus*],"[31] she says. Leakey freely admits limitations when it comes to assessing hominid fossils: "I'm no anatomist. I've just got a hunch." So the rationale for the anti-*Australopithecus* position is not well articulated, but it is nonetheless deeply felt. In fact, when the Hadar and Laetoli fossils were first discovered and interpreted as showing *Homo* and *Australo-pithecus* coexisting at 3 million years and more, Mary Leakey voiced her satisfaction at what this implied for the southern apes. "One of the most significant facts to emerge from Koobi Fora, Afar and Laetoli is that the australopithecines are really out on a limb,"[32] she wrote to a colleague at the end of 1975.

Knowing all this was one reason Johanson went through his and White's rationale in so much detail in his 23 December 1977 letter to Mary Leakey. "I knew I had to make a strong case,"[33] Johanson now says. Leakey's response was somewhat as Johanson and White had expected, but nevertheless was encouraging. Although in her reply to Johanson on 9 January 1978 Leakey noted that several "highly respected persons" had suggested that a new species name might be a good idea, and that she was "somewhat in agreement" with the idea, she cautioned that "the difficulty of arriving at a correct label has seemed to me insurmountable." The reason is made abundantly clear. "I do not think *Australopithecus* is correct. It is a lousy term, based on a juvenile for which there is doubt as to whether it is *A. africanus* or *A. robustus*. Nor is it a direct ancestor of *Homo*, as all of us people agree."

Although this seems to be a plain statement of disagreement, Leakey went on to offer a possible solution. "Essentially I agree that the Laetoli/Hadar creature must be named by you and Tim before one of our so-called 'colleagues' steps in. If you can find an alternative to *Australopithecus* or make it clear that I disagree on that issue . . . I'll join with you and Tim, otherwise I must stand down."

When Leakey's letter arrived in Cleveland, Johanson immediately got on the telephone to White, and during a long discussion they composed a response, which was mailed on 4 February. This new letter essentially reiterated the reasons that there was no logical alternative to *Australopithecus*, but also explained how their plan of publication might be acceptable to Leakey. "The scheme is two-fold. First, we simply name a new taxon in the proper manner and have it published as soon as possible, then Tim and I will do a joint paper on the phylogenetic implications etc." In this way, reasoned Johanson and White, Mary Leakey would be a coauthor in the naming paper only and would not have to be associated with a new family tree that made *Australopithecus afarensis* ancestral to *Homo*. Johanson closed his letter by saying that the three of them could discuss these issues when she came to the United States in March, just a month away. White wrote to Leakey on the same day, with a similar message.

"At this point, rather than engaging in a long correspondence we simply decided to wait until we could go over the details in person with Mary,"[34] explains White. "We especially wanted to show her the [First Family] cranial material which she was not aware of but which was so important in our views on the need for a new species name." That cranial material, which White and his colleague Bill Kimbel later assembled into a composite skull, showed the curious combination of a primitive, apelike skull with jaws that were undoubtedly hominid. This, Johanson and White hoped, would substitute for the cranial material that Mary Leakey had wished might be recovered from Laetoli.

For Johanson, the experience of the previous twelve months had been salutary. He had been made aware just how much his preconceptions had influenced his statements on his own fossils and on the shape of human evolution in these crucial early stages. "Yes, I was guilty of personal prejudices and beliefs,"[35] he now admits. "I was trying to jam the evidence of dates into a pattern that would support conclusions about fossils which, on closer inspection, the fossils themselves would not sustain."[36] The reason Johanson did this? "I 'knew' there were two kinds of hominid at the South African sites; two kinds at Olduvai; at least two kinds at Koobi Fora. So, there had to be two kinds at Hadar, perhaps even three. I think I was influenced by Richard. We were friends, and I wanted to sustain that. I was willing to listen to him and to Mary and say, yes, we have two kinds at Hadar too."[37]

The association with White had smashed this particular lens through which Johanson had viewed human evolution, and had substituted another. When he wrote to Mary Leakey at the beginning of 1978, suggesting they meet and talk about *afarensis* in March, Johanson was hoping to persuade her at least to acknowledge this new view and accept just a limited part of it. Johanson's success in this would be crucial to the reception of *afarensis*.

CHAPTER 12

LUCY: THE RESPONSE

E arly on the morning of 22 August 1978, Mary Leakey set out on the one-hour drive from her camp at Olduvai Gorge to the mist-shrouded town of Ngorongoro, named for the volcanic crater to whose upper slopes it clings. Leakey went straight to the town's post office and dispatched the following message: "JOHANSON NATURAL HISTORY MUSEUM WADE OVAL CLEVELAND USA PLEASE OMIT MY NAME FROM PAPER ON NEW SPECIES REGARDS MARY."

The telegram arrived at its destination a few hours later, but Johanson was not in the museum. "I was on my way to Sweden," he recalls, "where I was to be best man at Jack Harris' wedding."[1] The telegram was opened by Bill Kimbel, Johanson's colleague at the museum. Kimbel immediately had Johanson paged at New York's Kennedy Airport, where he was about to board his flight to Europe. Johanson rushed to a telephone to hear Bill saying, "What are we going to do? Mary Leakey has cabled saying she doesn't want her name on the paper. The whole thing is printed and ready to go." Thinking quickly, but with no real option, Johanson replied, "Don't let a single copy leave the museum. Get the title page reprinted without her name and the paper rebound. I don't want to get her upset."

That done, Johanson left for the wedding perplexed. During the long transatlantic flight Johanson's thoughts kept returning to

Mary Leakey's message. What had prompted her action? Hadn't it all been resolved and agreed during their discussions in March, when Mary was on her U.S. tour?

The paper in question was publication number 28 of the Cleveland Museum, entitled *Kirtlandia,* which contained the formal naming of *Australopithecus afarensis.* When Johanson had packed the manuscript off to the printer in the early summer, its authors had read as Donald C. Johanson, Tim D. White, Mary D. Leakey, and Yves Coppens, in that order. Coppens's name was listed primarily because of an agreement between the co-leaders of the Afar project made in 1972: this gave the Frenchman the option of appearing as a coauthor on major publications. In this case, as in others, he exercised it. Mary Leakey's name was there because, as the Laetoli fossils were part of the new species, it was only proper to include their discoverer. Johanson and White contend that she agreed to the arrangement in March. Leakey, however, insists that "They put my name on the paper without my permission."[2]

What actually happened is impossible to document because, unlike the discussions between Johanson, White, and Leakey during November 1977 through February 1978, which were conducted by letter, none of the exchanges that took place while Leakey was in the United States in March 1978 were committed to paper. In the normal course of events, there was no reason that they should have been. As it happened, however, when Leakey withdrew her name from the printed paper there followed strong contradictory assertions about what arrangement—if any—had been agreed to. This immediately provoked a swirl of speculation on each side of the issue as to the motivations of the other. Johanson and White found themselves accused of poor scholarship and of manipulating the system in search of extra kudos. Leakey is charged with sticking unquestioningly with demonstrably false scientific assertions.

The story therefore begins—or rather, continues—when Mary Leakey arrived in Berkeley, 28 February 1978. The following day she met White at 9 A.M. in his office in the anthropology department, and after sharing the usual pleasantries, turned to the subject of *afarensis.* White remembers the conversation this way:[3]

"Did you get Don's letter?" he asked.

"Yes, and I don't like it."

"Why don't you like it?"

"It's the use of the term *Australopithecus*—I hate the word."

"Why?"

"I don't know. I just don't like it."

White then began to go over again the rationale for calling the Laetoli/Hadar hominid *Australopithecus*.

"There are three choices," he told Leakey. "You can call it *Homo*, in which case you are putting a creature that is more primitive than any other hominid in the same genus as ourselves. You can name a new genus, but then you would have to explain why all these other things are so similar and yet are in a different genus. Or you can call it *Australopithecus* as we suggest, and retain a thread of logic in it. Those are the rules of nomenclature."[4] In other words, any new name proposed for the Laetoli/Hadar fossils had to be consistent with what had gone before. It had to reflect the primitiveness of the fossils and show how they related to the two hominid genera—*Australopithecus* and *Homo*—already known to exist at a later date. Very simply, *Australopithecus* was the closest model.

According to White, Leakey saw the force of the logic, and accepted that *Australopithecus* was the right choice. "I convinced her that *A. afarensis* was the taxonomic way to go,"[5] White noted in his diary that evening. At the end of their morning discussion, White placed a call to Johanson, who spoke at length with Leakey. According to Johanson, the conversation went as follows:

"All right," said Leakey. "I accept what you say, but I don't want to be associated with a phylogeny that puts *Australopithecus* as our ancestor."

"There are two steps to this," Johanson replied, reiterating what he'd written in his 4 February letter to Leakey. "The first is merely naming this new species in a straightforward way. And the second is discussing the relationships, the phylogeny, which Tim and I will do in a separate paper."[6]

Leakey is said by Johanson to have agreed to this arrangement, even though a close association with the naming part would surely have implied at least a brushing association with the phylogeny.

Mary Leakey's account of events is different. "I discussed the Hadar and Laetoli fossils with them, and I am quite sure that I went along with the idea that the bigger things at Hadar were very similar to those at Laetoli. But I always thought Lucy was different. I also disagreed with their interpretations. I said I didn't think the Laetoli specimens were *Australopithecus*. I objected to the term. That has been consistently my view."[7]

Whatever actually happened during those few hours' discussion in White's office, there followed that evening a very jovial dinner arranged for Leakey by archeologist Desmond Clark, and attended by White; Yves Coppens, who happened to be in town; and the late Glynn Isaac. Mary Leakey's afternoon seminar on the Laetoli footprints had gone well, and she was in her usual sharp form, taking pleasure in taunting Coppens. Later in the week Leakey continued on her U.S. tour, speaking about the Laetoli footprint trail in several cities and calling in briefly at Cleveland, where for the first time she was able to see the entire Hadar fossil collection.

Meanwhile, Johanson and White began the job of polishing the two planned manuscripts, the naming paper for *Kirtlandia* and the phylogeny paper, which was to be submitted to *Science*. At the end of April, Johanson took the precaution of sending a copy of the *Kirtlandia* manuscript to one of the world's great experts in evolutionary biology and systematics, Ernst Mayr, at Harvard. Johanson wanted to ensure that he and White had not unwittingly committed a taxonomic blunder that might invalidate their choice of a name for the Laetoli/Hadar hominid. On 4 May, Johanson received a short note from Mayr saying, "The manuscript as it is now before me would certainly legitimate your new name." Thus reassured, Johanson packed his bags and set out for Sweden, where he would speak at the Nobel Symposium, now fully prepared to pronounce for the first time in public the words *Australopithecus afarensis.*

The Nobel Symposium was to be a grand affair, organized by the Royal Swedish Academy of Sciences and attended midway through by the King and Queen of Sweden. As part of the ceremonies, which were separate from the scientific program, King Gustav was to present Mary Leakey with the Golden Linnaean Medal, in recognition of her contributions to the biological sciences. Leakey was to be the first woman to receive this high honor. The symposium itself was spread over six days and dealt with archeology as well as fossil hominids. As would befit such an occasion, the program would take a broad view of paleoanthropology.

The symposium had been some years in the planning, with Richard Leakey as prime mover. And one of his principal motivations was to attempt to improve the image of paleoanthropology. "I had been involved in trying to raise international funding for paleoanthropological studies, partly in relation to the Louis Leakey Memorial Institute, partly in relation to FROM—scholarship money, endowment money, and so on,"[8] he explains. "One kept running into the idea that paleoanthropology was not a science, and this sometimes made fund-raising difficult."

It is certainly true that in the spectrum of the sciences, from ("hard") physics to ("soft") biology, human evolutionary studies are usually regarded as being extremely "soft." The relative under-representation of paleoanthropology in Britain's Royal Society and the United States' National Academy of Sciences attests to this attitude. "It seemed to me," says Leakey, "that if we could get paleoanthropology recognized and talked about in connection with the Royal Swedish Academy of Sciences and the Nobel Foundation, many people in the United States and Britain might be willing to place paleoanthropology into another category, a more scientific category." Professional and social connections between Leakey and Professor Carl Gustav Bernhard, Secretary General of the Royal Swedish Academy of Sciences, extended back several years, and so the plan was not as difficult to implement as it might otherwise have been.

The result of the plan? "The aim of changing the image of paleoanthropology worked to a certain extent, and would have achieved more had it not been for the disaffection generated through the reaction to Don's book, Lucy, which was published in 1981. I wouldn't blame Don entirely for that," says Leakey. He blames the press for making "far too much" of the differences between him and Johanson.

Nevertheless, those differences appear to have been seeded at the Nobel Symposium. In retrospect, Leakey suspects that Johanson chose the occasion to announce his new species so as to "add the luster of Nobel to the naming."[9] Johanson counters by suggesting that Leakey's real motivation in organizing the meeting was to prepare the way for a Nobel Prize for himself. "Yves Coppens told me that by naming afarensis at the meeting I overshadowed Richard's work, which is why he was annoyed with me,"[10] says Johanson. Nobel Prizes in science are restricted to chemistry, physics, and medicine and do not include paleoanthropology. "The Nobel Foundation has nothing to do with giving Nobel Prizes,"[11] says Leakey. "There is no possibility of anyone getting a Nobel Prize in paleoanthropology."

The real reason that Leakey was annoyed with Johanson, he says, is that the Nobel Symposium was an inappropriate forum at which to name a new hominid species. "The conference wasn't of that kind," he insists. "It was to look at the interface between archeology and anthropology." Richard had known about the *Kirtlandia* paper, because Mary had earlier shown him, in confidence, a draft copy which Johanson had sent her. But he did not know that Johanson would choose to make the species announcement at the Nobel meeting. "The naming of *afarensis* was simply out of context," says Leakey. Johanson argues that the naming of a species was fully in keeping with the nature of the meeting, and points out that Phillip Tobias had in mind to name two subspecies of *Australopithecus* during its proceedings.

Mary Leakey did know that Johanson would name *afarensis* at the meeting, but was unaware that he would spend so much time talking about her site and her fossils. And although she had seen a draft copy of the *Kirtlandia* manuscript, she says she did not know that one of her Laetoli fossils would be designated as type specimen. As recounted in the previous chapter, this put her in a difficult position in speaking about Laetoli after Johanson's paper. But it seems possible that Mary's unease may have been exacerbated by Richard's comments when she showed him the *Kirtlandia* paper a few weeks earlier. "It's up to you, but I don't think it's one species," he had said. However, such a difference of professional opinion between Richard and Mary was nothing novel at that point. "We were fighting like mad at the time,"[12] remembers Mary Leakey. "We were disagreeing about virtually everything to do with human evolution. Interference from Richard would be more likely to push me in the other direction."

"Perhaps it was the cool reception given to *afarensis* at the Nobel Symposium that caused Mary to begin to break with Tim and Don,"[13] offers Richard Leakey. Whatever was the case, by midsummer Mary Leakey was distinctly angered by events, so much so that by the time White arrived at the Laetoli camp on 4 July she had already been voicing her annoyance with what had taken place in Sweden in May. And very soon she and White began to argue about a number of issues too. "White was obnoxious that season,"[14] recalls Leakey. "He wanted to direct everything that was going on. He wouldn't give the men credit for anything. He wanted to take over the digging. I think he resented the fact that I was boss." Previously White and Leakey had experienced a very good and close relationship, in both professional and social dimensions. But on this, his first field season at Laetoli, White had quickly formed a low opinion of the work going on there, and had tried to do something about it, perhaps with rather less tact than he should have applied. Hence the friction.

But the real breaking point was over the fossils themselves. Mary Leakey was saying that she did not agree that the Laetoli/Hadar hominids were all one species. And she certainly did not believe them to be *Australopithecus*. Indeed, at one point after White had left the camp she said to others still there, "Call it what you like, call it *Hylobates*, call it *Sympholangus*, call it anything, but don't call it *Australopithecus*."

While he was in camp, White had tried to argue again with Leakey as to why *Australopithecus* was the only logical choice, and the exchanges became quite heated. "I had to listen in the workroom at my own camp at Laetoli to a long harangue from Tim White in an attempt to get me to change my mind,"[15] Leakey commented later. At one point, on 21 August, White exploded, "Damn it, Mary, your name is on the paper and I don't want to hear about it when it's published. If you are really against our interpretation, take your name off."[16] Leakey sometimes says that it wasn't until this point that she actually knew her name was on the *Kirtlandia* paper. In any case, the following day she made the journey to Ngorongoro, cabled Johanson, and asked that her name be removed.

Two weeks later White left the Laetoli camp, never to return while Leakey was there.

The *Kirtlandia* paper was eventually published toward the end of the year, its title page recast and the whole document rebound. Because the journal is one of the more obscure backwaters of the scientific literature, its appearance did not occasion any public comment. That would have to await the publication of the *Science* paper, which came out on 29 January 1979. Johanson arranged for a press conference at the Cleveland Museum to coincide with publication, at which he and White described their fossils and explained why a completely new view of human evolution was required as a result of their work. The human family tree was a simple Y-shape, argued Johanson and White: *Australopithecus afarensis* formed the vertical stem, with more advanced australopithecines taking one branch to an eventual dead end, and the other the line leading to ourselves. It was a simple, powerful argument. And it was an important moment for the museum and for Johanson.

Once again, *Nature's* anonymous Palaeoanthropology Correspondent took the opportunity to comment. Exactly four years earlier, a correspondent in the same columns had been extremely sniffy about the way Johanson and his colleagues had announced their findings from the field, without first completing detailed analysis: "Such a reversal of priorities, with its consequent and inevitable de-emphasis on necessary laboratory procedures is completely inconsistent with current scientific method and theory."[17] Now that the analysis had been done—four years of it—the correspondent remained equally unimpressed. "[Johanson and White's] success in demonstrating the morphological distinctiveness is by no means clear,"[18] the correspondent opined. In an apparent display of where his or her sympathies lay, the correspondent added: "It should be pointed out that in the original report on the Hadar material, specimens showing these features were placed not in a primitive hominid group as they now are, but in a more advanced genus, *Homo.*"

Critical though this was, the real battle of the bones, as Johanson has called it, took place in the pages of the more popular press. *The New York Times* of 18 February set the confrontational tone, its report accompanied by the front-page photograph of Leakey and Johanson apparently debating their differences.

Leakey argues that the "confrontation" described in the *Times* was more apparent than real, a piece of manufactured news. Perhaps that is what news editors were seeking. In any case, a report in *Newsweek* at about the same time went out of its way to seek Leakey's contrary opinion to Johanson's claims. "I am perfectly prepared to accept that Lucy is a new species of *Australopithecus,*"[19] Leakey is reported to have said. "But I believe that the common ancestor has not yet been found." Generally, however, the *Newsweek* report was positive about Johanson and White's paper, unlike its counterpart, *Time.* "Johanson's announcement ... left most of his colleagues puzzled,"[20] ventured *Time.* "The bones have been around for more than four years now, long since dated by potassium/argon tests, and many anthropologists who have studied them are generally convinced that Lucy is *Australopithecus africanus,* not some new species at all." Johanson wondered whether Leakey's influence was in some way behind *Time's* negative attitude.

Exchanges of this kind are perhaps to be expected, given the nature of the subject. But the tone sharpened distinctly when, a few weeks later, Mary Leakey was again in the United States describing further progress with the Laetoli footprints. At a National Geographic press conference in Washington, D.C., she was asked her views on *afarensis*. At first she declined to comment, but when pressed said she considered Johanson and White's work "not very scientific."[21] Offered an opportunity to respond, Johanson claimed that Mary Leakey "really shows a poor appreciation of what evolution is all about." Clearly, there was a good deal of bitterness behind both sets of remarks.

Mary Leakey's pithy barb was not without substance. A set of criticisms of the naming of *afarensis* was raised that persist to this day. These relate to the motives and implications of choosing Laetoli hominid 4 as the type specimen of *afarensis,* and the wisdom of uniting two sets of fossil hominids separated by 1,500 kilometers and half a million years under the same species name.

"In the *Kirtlandia* paper we state quite clearly why we chose LH-4 as the type specimen,"[22] explains White. "One, it was already published, in the *Nature* paper I did with Mary and others in 1976. And you won't find better descriptions in the literature than that. Two, it is diagnostic. No one has ever questioned that. It makes a perfectly good type, and everyone knows it. Three, it has served to make known the connection between the Laetoli and Hadar samples of hominids." It should be said that excruciatingly detailed though the International Code of Zoological Nomenclature is, there are no strict rules about the choice of type specimen, so long as other fossils can usefully be compared with it. Specifically, there is no requirement that the chosen type specimen should have been described in a previously published paper.

But though Johanson and White broke no rules in designating the Laetoli fossil as type specimen, there was nevertheless a near-universal outcry at the choice.

Mary Leakey, for instance, thinks that the choice was made not so much to "make known the connection between Hadar and Laetoli" as to associate *afarensis* with the secure, older date of Laetoli. "This would make their species the oldest yet known," she says. "And they wanted the authority of my name."[23] Leakey is not alone in suspecting such motives. "Their stated reason for choosing LH-4—that it was already published—is pretty hollow,"[24] agrees Michael Day. "It was important to them because it gave them a very old date, the oldest hominid in the world."

Many anthropologists argue that the logical choice for type specimen would have been Lucy, simply because she represents a wide range of body parts in her skeleton, as against the single, worn lower-jaw fragment that is LH-4. Ernst Mayr's comment crystallizes this sentiment: "It is so obvious to choose the Lucy skeleton. There is so much more of it than this scrap of jaw from Laetoli."[25] Mary Leakey's view was clear: "It is regrettable . . . that the type specimen selected should be a worn mandible from Laetoli, when much better-preserved specimens are available from the Afar itself,"[26] she said. "Lucy would have been the obvious choice."[27]

The argument that Lucy's partial skeleton provides so many more bones against which other bones can be compared certainly sounds persuasive. To which Johanson responds as follows: "Why not Lucy? Because the basic distinction between *afarensis* and other australopithecines is in the teeth and cranial anatomy. Lucy's ribs and vertebrae, for instance, aren't distinctive. They don't distinguish her from any other species of *Australopithecus*. It is like a steering wheel and four wheels on a car: by themselves they don't distinguish between a Ford and Toyota. You have to have characteristics that do distinguish. The dentition of LH-4 does this perfectly for *afarensis*. We could have erected a new species on this specimen alone."[28]

White points out that Lucy is part of the formal naming process anyway (as one of the so-called paratypes), even if she isn't the type specimen. Her skeleton is therefore formally available for any anatomical comparisons that need to be done. "A lot of people in this field make a big deal about the choice of a type specimen," he says. "In fact, it is one of the least significant aspects of the whole issue."[29] He suspects that at least some of the intensity of criticism directed at this issue is simply a sublimation of more general anti-*afarensis* or even anti-Johanson/White sentiments.

A second, more serious, criticism is the choice of the name *afarensis* itself, which was selected to honor the area of Ethiopia from which the major fossil collection was recovered. But, again, it seems possible that some of those who pursue the criticism may well be motivated by secondary—perhaps unconscious—agenda.

The problem here stems from the possibility that the hominids living at Laetoli might be different in some degree from those located 1,500 kilometers to the north at Hadar and half a million years later in time. Geneticists know that modern populations of a species that are separated by even modest geographical barriers can have distinct genetic profiles and may even have a slightly different appearance: they would be geographic variants, or even subspecies. There is therefore a strong possibility that the separation by half a continent and half a million years between the Laetoli and Hadar hominids would have engendered some important differences, even if these are not readily identifiable in the fossil fragments so far available.

So how does this make the name *afarensis* a problem? "You cannot have things that are in totally different localities and different times, and then pick the name from one place [the Afar] and the type specimen from the other [Laetoli]," says Ernst Mayr. "If you select a geographical locality for the name, then you have no choice but to select the type specimen from the same locality."[30] And the reason for this is that if by some means it should be discovered that the Hadar hominids are in fact different and deserve a different name from those at Laetoli, then, because of the rules of nomenclature, the Laetoli hominids must continue to be called *afarensis* while those in the Afar would be named something else. This would be confusing, to say the least. The rules of zoological nomenclature are governed by constancy, not logic: once something is named it is virtually impossible to "unname" it, no matter how illogical the result might sound.

If Johanson and White had called their species *johansonensis*, then the problem Mayr raises would not occur. The point of contention lies in choosing the name of one geographical area and applying it to another, Afar to Laetoli.

In correspondence with Phillip Tobias about *afarensis*, Mayr felt so strongly about this potential problem that he said, "I urge students of fossil man to ask the Commission to suppress Johanson's type designation and designate instead a specimen from Afar as type."[31] Mayr explained his motivation by saying, "It cannot be emphasized often enough that every population is variable but also that every species consists of numerous local populations differing to the degree of their isolation. This of course makes things complicated but the fact of the matter is that evolution is complicated." Mayr has no professional vested interest in the applicability of this argument to the proper naming of the Laetoli/Hadar hominids, but his friend Phillip Tobias does. It so happened that Tobias had already formed the opinion that the Laetoli and Hadar hominids were indeed subspecies, but of *A. africanus*, not *A. afarensis*, and this was what he had planned to announce at the Nobel Symposium when Johanson scooped him. Tobias proposes that the Hadar creatures should be called *Australopithecus africanus aethiopicus* and those from Laetoli *Australopithecus africanus tanzaniensis*, which is the jaw-splitting way of designating subspecies. As a corollary, he argued at an international scientific meeting in London in March 1980 that: "Since the tying of the name '*A. afarensis*' to the Laetoli fossils is manifestly inappropriate and since it is considered that the case for '*A. afarensis*' has not been established, it is proposed formally that the name '*A. afarensis*' be suppressed."[32] Tobias has repeated his proposal at several important scientific gatherings in recent years, but *A. afarensis* remains intact and usually not encaged in denigratory quotation marks.

Tobias occupies the chair of anatomy at the University of the Witwatersrand in Johannesburg, the chair previously held by Raymond Dart, discoverer of the first *Australopithecus africanus* fossil, the Taung child. This species of *Australopithecus* has for two decades been the most widely accepted candidate for the rootstock of the *Homo* lineage. It is therefore perhaps not surprising that as official custodian of the Taung fossil, Tobias also acts as de facto protector of the ancestral status of *Australopithecus africanus*. Although he once published a human family tree showing a hypothetical species, *Australopithecus* "?," as precursor to *africanus* and *Homo*, he no longer favors the idea—not in any concrete terms, at least. Johanson, not surprisingly, says "I believe that this postulated australopithecine is indeed *Australopithecus afarensis*."[33]

Given Tobias' opinion of *afarensis*, he was naturally delighted to receive Mayr's suggestion that something should be done to change the the type specimen of the species. "I am gratified indeed that so eminent an authority on taxonomy as yourself is able to agree in large measure with my conclusions concerning *Australopithecus afarensis*,"[34] he replied. In fact, the level of agreement between the two men lies more in the arena of correct nomenclature than in interpretation of the fossils themselves. "I accept *Australopithecus afarensis* as a new hominid type," says Mayr. "It is not very different from *africanus*, but it is different."[35] Nevertheless, if the International Commission on Zoological Nomenclature were simply to force a change in designation of type specimen, it would represent a small victory at least for Tobias. "How should one proceed?" he asked Mayr.

Mayr explained that a simple letter to, for instance, *Nature* would not be sufficient, as Tobias had hoped it might. Instead, the two men must apply formally to the Commission and ask that Laetoli hominid 4 be suppressed as the type specimen and that Lucy be designated as the new type specimen, or "neotype." "It is perfectly legal for the two of us to send an application to the Commission," explained Mayr. "However it might lend more weight to the application if it were also cosigned by others including, for instance, some society of anthropologists."[36] This is as far as the plan proceeded, however, principally because Tobias became diverted by other events, not least of which was the increasing battle against apartheid in South Africa, in which he is deeply and courageously involved.

Mary Leakey also promulgates this line of criticism, and this was the basis of her remark that Johanson and White had been "not very scientific." At an international scientific meeting in London, Leakey described their action as "the arbitrary application of the same specific name to the hominids from the two localities, which are separated by over 1000 miles."[37] Johanson's response is to remind paleoanthropologists that traditionally, geography and time have not been considered important in the identification of relatedness—anatomy should be the guide, he says. "And yet some people who criticize us [on afarensis] . . . would recognize Homo erectus at 1.5 million years in East Africa and maybe at half a million years in China. It's the anatomy that draws fossils together in one species."[38] This was the basis of Johanson's remark that Mary Leakey "really shows a poor appreciation of what evolution is all about."

The third line of assault on afarensis came from deep within the Code of Nomenclature itself, and was the subject of a long letter to Science by Mary Leakey, Michael Day, and Todd Olson. "In this comment it is our purpose to establish the relationship between Meganthropus africanus Weinert 1950 and Johanson's replacement name for the taxon, as well as discuss the errors that we believe have been incorporated by Johanson and White into their taxonomic speculations,"[39] they wrote in March 1980. "Hairsplitting obfuscation"[40] is how Johanson characterized—and dismissed—the challenge.

The Leakey/Day/Olson argument had to do with the fact that a German scientist, Ludwig Kohl-Larsen, had found a piece of hominid lower jaw at Laetoli in 1939, which was later given the name Meganthropus africanus by Hans Weinert. When Johanson and White pooled the Hadar and Laetoli fossils in naming them Aus-

tralopithecus afarensis, they included this little jaw. According to the code, therefore, the Laetoli/Hadar fossil hominid should be called *Meganthropus africanus,* not *Australopithecus afarensis,* or anything else, for that matter. Had Johanson and White blundered after all?

Finding an anthropologist who both understands the legalistic interstices of the code sufficiently to pass an authoritative opinion and is also unbiased in the *afarensis* debate is not easy. But British anthropologist Bernard Campbell is a good candidate. In the 1960s he had helped rationalize the great menagerie of fossil-hominid names, culling a list of more than a hundred down to a small handful. He had also attacked the validity of *Homo habilis* when it was first proposed, but later relented.

On *afarensis* Campbell has this to say: "It would of course be valid to call it *Meganthropus africanus,* but this doesn't seem logical. The fossils are so like existing *Australopithecus* that they should clearly be placed in this genus. Now, because you are supposed to keep the species name constant, this would make them *Australopithecus africanus.* But you can't do that, because this species name is already 'occupied,' and it's known. So now you are free to create a new name. I think the new name, *Australopithecus afarensis,* is valid."[41]

Mary Leakey had initiated the line of criticism presented in the letter to *Science,* specifically by asking Day if he would explore the nomenclatural issues that might possibly have ensnared Johanson and White. Day at this point had consulted Campbell, who gave his opinion that the name was indeed valid. Nevertheless, Day felt compelled to continue and eventually drafted the letter to *Science,* in collaboration with Todd Olson.

The *Science* letter, however, has had little impact within the profession. "*Afarensis* is now in general use," says Campbell, "which is another point in its favor. As a label it is therefore very useful."

Meanwhile, Mary Leakey wonders whether a second nomenclatural approach might be more successful. A number of researchers consider that the Laetoli/Hadar fossils are very similar to, if not identical with, fossils from one of the South African sites, at Makapansgaat in the northern Transvaal. These fossils were once given the name *Australopithecus prometheus,* because it was thought— mistakenly, as it turned out—that they had used fire. If it can be shown that the Laetoli/Hadar hominids were indeed identical with the Makapansgaat hominids, then, following the rules of the Code, *Australopithecus afarensis* would have to be suppressed and be replaced by *Australopithecus prometheus*. Mary Leakey has asked Phillip Tobias to pursue this possibility, so far without result.

And so it goes.

On the central issue of the Laetoli/Hadar fossils themselves, the paleoanthropological community remains divided. Do the fossils represent just one species, which then may or may not be ancestral to all later hominids? Or are they a mixture of two species, or even more? The spectrum of opinions is interesting, and is well illustrated by the interpretation of the postcranial bones, specifically the limbs.

At one end—in support of *afarensis*—is Owen Lovejoy, an anatomist at Kent State University and close colleague of Johanson and White. He believes that the fossils belong to one species, that the anatomical variation in the fossils is one of size only, and that the big ones (males) moved around in the same manner as the small ones (females).

In the middle of the spectrum is a group of researchers at the State University of New York at Stony Brook, who include Randall Sussman, Jack Stern, and Bill Jungers. From their analysis of the limb anatomy, they conclude that yes, the bones probably belong to one species. But they also consider that there is sufficient difference in the anatomy between the big ones (males) and the small ones (females) that their patterns of locomotion would be different: both would be essentially bipedal, much like modern humans, but the females would have spent more time than the males climbing in the trees. This latter pattern of behavior would be similar to that of modern orangutans.

At the other end of the spectrum are Yves Coppens and his colleagues in Paris. Although Coppens was one of the coauthors of the original naming paper in *Kirtlandia*, he now believes, on the basis of anatomical differences in the arm and leg bones, that two species lived at the Hadar, and maybe more. He would continue to call one of them *Australopithecus afarensis*, and this would include Lucy and the other diminutive individuals. But, he says, there is a primitive species of *Homo* present too.

So, after analyzing the same set of fossils, three different research groups come to three different conclusions. "No position is overwhelmingly strong," observes David Pilbeam, "which probably means there simply isn't enough fossil material available to allow a fully objective assessment."[42] He also points out that if paleoanthropologists are truly trying to look back in time to a point from which two hominid lines diverged, then the closer they come to that point, the more difficult it will be to distinguish the members of one newly emerging line from those of the other, simply because the two inevitably will look very much like each other. It is precisely under such circumstances—when objective uncertainty inexorably rises—that subjective preconceptions can exert their greatest effect.

Both Johanson and White recognize this in Richard and Mary Leakey's reaction to *afarensis*. "If you want to know the real reason why the Leakeys are upset," offers White, "it is because we took away the option of saying the Leakeys' hominids are early, true *Homo*."[43] At the same time, White contends that his own work is based on the facts, not preconceptions.

Richard Leakey's response is the following: "All I've ever said is that my preference would be for another interpretation. I couldn't rule out the possibility that they are right and I don't think they can rule out the possibility that they are wrong."[44] He does suggest White may be unconsciously biased. "I think the single-species overprint on Tim is much greater than he realizes. He's not aware of the fact that he's prejudiced. Inevitably, he must have been influenced by his experience at Michigan, so that he fails to see what so many other people see in varying degrees of clarity." Of his own preconceptions, Leakey says they are simply that a proper study of the fossils will eventually reveal the truth.

All of which suggests that it is easier to recognize bias in others than to admit it in oneself. It also probably means that some questions in paleoanthropology may well be impossible to answer with any degree of certainty—and human beings dislike uncertainty, especially when it concerns themselves. Combine these two truths and you get an inevitable result, as noted by Johanson: "Anthropologists who deal with human fossils tend to get very emotionally involved with their bones."[45]

CHAPTER 13

MAN'S PLACE IN NATURE

"What is the role and status of our own species, *Homo sapiens*, in nature and the cosmos?"[1] This, suggests Stephen Jay Gould, of Harvard University, is the "cardinal question of intellectual history."

Certainly, it seems to be an old question, having etched itself into the philosophy of virtually every society for which there are records. Without doubt, it is a controversial question, at least in the manner with which it has been grappled within the materialistic realm of Western philosophy. As Smithsonian Museum scientist Gerrit Miller observed in 1928: "Among recent subjects of animated scientific and popular controversy both in and out of print there is perhaps none that has aroused more widespread interest than the discussion of human 'missing links.' "[2] And to judge by subsequent history, some of which has been recounted in the pages of this book, nothing much has changed since Miller's time.

Is the cardinal question perhaps, by its very nature, simply unsolvable—even, or maybe particularly not, by the methodology of objective, scientific investigation?

Duke University anthropologist Matt Cartmill has this to say on the subject in general and his colleagues' scientific discipline in particular: "The demands of the scientific method itself force us to pursue the essentially extrascientific objective of telling stories that explain our privileged status in the universe of things."[3] Cartmill's audience for these remarks was a gathering of physical anthropologists at one of their society's recent annual meetings: they were clearly interested in the ideas, but, perhaps not unnaturally, seemed to wish to hold them at arm's length. "The importance of our science lies in its effects on our world view—on the way people think of themselves and the universe and their place in it—which is a subject within the province of ideology and religion, broadly defined," he added.

Cartmill told his audience that when he had published similar sentiments six months earlier, he had received a number of letters from paleoanthropologists who "were not very pleased." Apparently, he said, they felt that "I had slandered our profession by questioning its purity, by depicting it as contaminated with unscientific elements." Not surprisingly, for no scientist likes to be told that what he is doing is "within the province of ideology and religion," however noble a pursuit it might be.

And, not surprisingly too, many contemporary anthropologists' reaction to all this is "Well, I guess in the early days people's work used to be affected by this sort of thing—ideology, mythology, and so on—but not now; not now that anthropology is *really* scientific." Cartmill's riposte is simple and direct: "This tendency to rescue scientific appearances by evading the mythological point of our science has distorted paleoanthropological thought through most of the twentieth century." This is a pointed challenge indeed, and needs to be set in perspective.

Twentieth-century paleoanthropology is, at its most fundamental levels, concerned with what Darwin's friend and colleague Thomas Henry Huxley characterized as "Man's Place in Nature," a subject on which he wrote a book in 1863. The subject, as Gerrit Miller articulated it, is simple: "Is man a creature unconnected with the rest of animate nature? Or is he a direct descendant from ancestors which were not human?"[4]

Before the theory of evolution became an established part of Western biological thought, humans were considered to be one of God's created creatures, but a very special one at that. We were seen as being uniquely endowed with very special capabilities and talents: possessed not only of a transcendent intelligence, but of moral and spiritual senses too. We were clearly different and separate from the rest of the animal world. We were, in Miller's words, a creature unconnected with the rest of animate nature. Naturally, with the acceptance of Darwin's concept of evolution, *Homo sapiens* perforce was viewed as a product of nature, just like all the other species on the planet. Like all the other species, and yet not so, because with our high intelligence, our moral and spiritual sense, we were clearly different, apparently both in degree and in kind. And this difference, this perceived gulf between *Homo sapiens* and the rest of animate nature, has been the focus and cause of many of the intellectual struggles in paleoanthropology.

It is ironic that this perceived gulf was a problem for pre- and post-evolutionary worldviews alike. The gap had to be explained —or explained away—and it is more than a little instructive that the methods by which it was achieved were common to both eras: investigators tended to see in the evidence what they expected to see in light of their theories.

For the pre-evolutionary world of science, the most dominant concept for establishing the order of things, including man's place in nature, was the Great Chain of Being. "The chain," explains Gould, "is a static ordering of unchanging, created entities—a set of creatures placed by God in fixed position of an ascending hierarchy that does not represent time or history, but the eternal order of things."[5] The concept had social as well as scientific force. "The ideological function of the chain is rooted in its static nature: each creature must be satisfied with its assigned place—the serf in his hovel as well as the lord in his castle—for any attempt to rise will disrupt the universe's established order."

The chain of being was therefore a descriptive and an explanatory device: it represented the world as it was perceived to be and clearly the way it ought to be. As the chain was the product of the Western European mind, it is not surprising that in the gradation of the "lowest" forms to the "highest," the European ideal was placed as its highest and most perfect terrestrial link—just "a little lower than the angels," to appropriate the description in Psalms. "Ascending the line of gradation, we come at last to the white European; who being most removed from the brute creation, may, on that account, be considered as the most beautiful of the human race,"[6] opined Charles White, a British physician who launched a major defense of the chain-of-being concept in 1799. White concluded a panegyric on the supposedly superior qualities of the European form with the following: "Where, except on the bosom of the European woman, [can one find] two such plump and snow white hemispheres, tipt with vermillian?" Quite so.

White's essay was, in the light of today's values, an outright racist document, albeit couched in language unusual for a scientific paper. "He was merely expressing a common opinion of his time in admittedly overblown rhetoric,"[7] notes Gould. As we shall see, racism in these terms has been a persistent theme of anthropology for a very long time.

As with all scientific theories, the chain of being had unexplained discontinuities—in this case quite literally so. Instead of its presenting a steady and continuous gradation throughout all of the natural world, there were large apparent gaps: namely, between minerals and plants; between plants and animals; and, most embarrassing of all, between apes and humans. So influential was the theory that when, in 1758, Carolus Linnaeus established the basis of zoological classification—his Systema Naturae—he postulated the existence of a primitive form of human, *Homo troglodytes*, that filled the chasm between humans and apes. *Homo troglodytes* was said to live in forests, to be active only at night, and to communicate only in hisses. At a time when the early African explorations frequently returned with fantastic tales of half-ape, half-human creatures, anything was possible, especially if theory demanded it.

Responding to these same demands of theory, Edward Tyson had earlier, in 1699, unconsciously manipulated what was to be the first scientific description of a great ape, in this case a juvenile chimpanzee. At the time, European scientists had only the haziest of notions about the monkeys, apes, and "primitive" peoples of the tropical Old World, and there was much confusion about the distinction, if any, between them. In describing the chimp, which he referred to as a Pygmie, he said, "Our Pygmie is no man, nor yet the common ape; but a sort of animal between both." Supporting this conclusion, Tyson sketched the animal in more or less fully upright stance, albeit with the aid of a walking stick in one case and straphanging from a rope in another. He had seen the animal "knuckle walk," as both chimpanzees and gorillas are now known to do, but assumed that it was an unnatural posture due to weakness following its long sea journey. "In this chain of creation, as an intermediate link between an ape and a man, I would place our pygmie," concluded Tyson.

Now, from the perspective of later years, when many anthropologists would vigorously deny a close relationship between humans and chimpanzees and would in recent times be shown to be wrong, Tyson's assessment looks surprisingly modern. But as Gould points out, "The outstanding feature of Tyson's treatise is not an accuracy born from casting aside old prejudices but rather Tyson's *exaggeration* of the human-like character of his pygmy . . . a result of his prior commitment to the chain of being." Tyson saw what he expected to see. "Intermediate forms were anticipated and expected, and Tyson's discovery produced a welcome confirmation of an established theory."[8]

In the post-Darwinian era, throughout the history of paleoanthropology, authorities would commit Tyson's error time and time again: Neanderthal, Piltdown, *Australopithecus, Ramapithecus, Zinjanthropus*—each in its turn has been the object of the exaggeration of traits favored by observers whose theories demanded them.

With continuing explorations of the "dark continent" through the eighteenth and nineteenth centuries, the apes began to be seen increasingly for what they were and technologically primitive peoples for what they were not. Shocked by what were interpreted as rudimentary cultures and societies, pre-Darwinian Europe assumed that here indeed were real gradations between the unquestionably supreme white race and brute nature. "Examples are not wanting of races placed so low that they quite naturally appeared to resemble the ape tribe," commented one French anthropologist of the time. Carl Vogt, a Geneva scholar of the mid nineteenth century, went into some detail: "The pendulous abdomen of the lower races . . . shows an approximation to the ape, as do also the want of calves, the flatness of the thighs, the pointed form of the buttocks and the leanness of the upper arm."[9] In their behavior, the "lower races" also matched what was expected of them. "Young orangs and chimpanzees are good natured, amiable, intelligent beings, very apt to learn and become civilized," he observed. "After [puberty] they are obstinate beasts, incapable of any improvement. And so it is with the Negro."

For good measure, Vogt displays his bias of male superiority. "We may be sure that wherever we perceive an approach to the animal type, the female is nearer to it than the male," he opines. "Hence we should discover a greater [apelike] resemblance if we were to take a female as our standard."

First in Europe and then in the United States, there developed through the eighteenth and nineteenth centuries clearly articulated and quantified assessment of the relative merits of the various races. For instance, the findings presented by Robert Dunn to the 1862 annual meeting of the British Association for the Advancement of Science were reported as follows: "He observed that the leading characters of the various races of mankind have been maintained to be simply representatives of a particular type in the development of the highest or Caucasian; the Negro exhibiting permanently the imperfect brow, projecting lower jaw, and slender bent limbs of the Caucasian child some considerable time before its birth, the aboriginal Americans representing the same nearer birth, and the Mongolian the same newly born."[10] At about the same time, America's leading zoologist, Louis Agassiz, noted that "The brain of the Negro is that of the imperfect brain of a seven-month infant in the womb of a white." The differences between the races were evident for all to see, and, in a pre-Darwinian era, marked a clear gradation from the lowest to the highest races in God's creation. God, in His wisdom, had placed Caucasians on the top of the heap.

One of the most celebrated anthropological cases of the mid-nineteenth century was that of Samuel George Morton, an American scientist and physician who earned a reputation for good and careful work. During the 1840s he published a series of papers on his measurements of the cranial capacities of several races, employing techniques he painstakingly developed himself. The results, observes Gould, "matched every good Yankee's prejudice —whites on top, Indians in the middle, and blacks on the bottom."[11] In a detailed analysis of Morton's work, Gould showed that the raw data simply did not substantiate the conclusions that had been drawn from them. "In short, and to put it bluntly, Morton's summaries are a patchwork of fudging and finagling in the clear interest of controlling a priori convictions." A damning accusation for any scientist. "Yet," says Gould, "I find no evidence of a conscious fraud; indeed, had Morton been a conscious fudger, he would not have published his data so openly." Morton, it seemed, had erroneously but subconsciously extracted from the data the pattern he expected—and wanted—to see.

Although Morton's data and conclusions were obtained and published in pre-Darwinian days, they were reprinted and used long after evolutionary theory became established. What had been seen previously as a gradation of races in God's creation was simply transformed into a gradation as a result of differential evolutionary success. The putative lower races were perceived in a sense as living fossils, and Morton's data were harnessed in support of this new set of ideas. Therefore, from the point of view of Western intellectual tradition, man's place in nature ultimately didn't change through the Darwinian revolution; it was simply that the means by which the Caucasian had achieved his evident superiority was acknowledged to be different, that it was the result of naturalistic rather than divine processes.

Racism, as we would characterize it today, was explicit in the writings of virtually all the major anthropologists of the first decades of this century, simply because it was the generally accepted world view. The language of the epic tale so often employed by Arthur Keith, Grafton Elliot Smith, Henry Fairfield Osborn, and their contemporaries fitted perfectly an imperialistic view of the world, in which Caucasians were the most revered product of a grand evolutionary march to nobility. Human progress through prehistory had, according to Keith, been "a glorious exodus leading to the domination of earth, sea and sky."[12] The same stirring tones are to be discerned in Osborn's championing of the high plateaus of Central Asia as site of man's origins, his "rise to Parnassus." It is not surprising, then, that these men interpreted the evidently dominant position of the Caucasian race as the natural product of the evolutionary process.

Roy Chapman Andrews, Osborn's close colleague at the American Museum, stated the issue bluntly. "The progress of the different races was unequal," he said. "Some developed into masters of the world at an incredible speed. But the Tasmanians, who became extinct about 1870, and the existing Australian aborigines lagged far behind . . . not much advanced beyond the stages of Neanderthal man."[13]

Keith judged the anatomical differences between the races to be large enough to have required a considerable time for their evolution. "A human type changes very slowly," he said. "Therefore we must make a liberal allowance for the mere differentiation of the modern type of man into distinct racial forms. . . . I do not think that any period less than the whole of the length of the Pleistocene period, even if we estimate its duration at half a million years, is more than sufficient to cover the time required for the differentiation and distribution for the modern races of mankind."[14] Many of the evolutionary trees published in books and papers by Keith and his colleagues reflected the supposed ancient origin of the modern races.

Such a long evolutionary separation—the Pleistocene is now calculated at 2 million years in duration—between the races would of course give ample opportunity for the cutting edge of competition to do its work. And Keith revered the stern and impartial judgment imposed by competition. "When we look at the world of men as it exists now, we see that certain races are becoming dominant; others are disappearing," he said. "Competition is not confined to human rivalries and struggles; it pervades the whole animal kingdom of life; it is the basis of Darwin's doctrine of evolution; it has been, and ever will be, the means of progressive evolution . . . To extinguish the spirit of competition is to seek racial suicide."[15] Osborn concurred: "The law of survival of the fittest is not a theory, but a fact."[16]

The fittest, it was generally agreed, could not possibly be races from the tropics because the tropics induced indolence and degeneration, not improvement. "The evolution of man is arrested or retrogressive . . . in tropical and semi-tropical regions," says Osborn, "where natural fruits abound and human effort—individual and racial—immediately ceases."[17] Clearly, without effort there is no improvement—a good Puritan ethic. Even Robert Broom, who worked for many years in Africa, agreed with this sentiment. "It seems impossible for the higher types of man even to live for any length of time in the tropics without degenerating," he wrote in 1933. "Apparently a steady improvement of the brain was only possible in a temperate climate."[18]

So it was that several threads of argument were woven together to form a theoretical fabric whose pattern matched closely the ethos of the Edwardian world. If the white races were economically and territorially dominant in the world, it was surely the natural outcome of natural processes. The slow pace of evolutionary change, the long separation between the races, the inimical environment of the tropics—all combined to produce a graded series of races, rising from the Australian aborigines at the bottom, through the black races and the Mongols, and reaching the Caucasians at the apex.

If man's place in nature appeared to be readily explicable and ordered through the races of *Homo sapiens* itself, by contrast it presented problems in the larger scheme of things. The perceived gulf between man and the brutes, though edged into by the "lower races," was still large. Thomas Henry Huxley commented upon it this way: "No one is more strongly convinced than I am of the vastness of the gulf between . . . man and the brutes . . . for, he alone possesses the marvellous endowment of intelligible and rational speech [and] . . . stands raised upon it as on a mountain top, far above the level of his humble fellows, and transfigured from his grosser nature by reflecting, here and there, a ray from the infinite source of truth." The origin of so important a gap demanded an explanation from evolutionary theory.

What in fact has happened through much of the course of paleo-anthropology is that practitioners have impaled themselves on the horns of a dilemma when approaching this challenge. On one hand, they have recognized that according to evolutionary theory, natural forces must be capable of essentially transforming an ape into a human. But on the other, they have until recently tended to concentrate on those characters which we feel make us special, such as intelligence, culture, society, and moral sense. "In accepting this persistently pre-Darwinian definition of their problem, scientists who study human evolution have saddled themselves with the paradoxical job of explaining how causes operating throughout nature have in the case of *Homo sapiens* produced an effect that is radically unlike anything else in nature,"[19] comment Matt Cartmill, David Pilbeam, and (the late) Glynn Isaac in a recent review. This "paradoxical job" is precisely what Cartmill was referring to earlier when he said, "The demands of the scientific method itself force us to pursue the essentially extrascientific objective of telling stories that explain our privileged status in the universe of things."

For some people, most notably Alfred Russel Wallace, who was the coinventor with Darwin of the theory of natural selection, and Robert Broom, the job simply proved to be too much, though in different ways. Both men concluded that human intelligence and moral sense had no other explanation than spiritual intervention.

As befits the coinventor of the theory of natural selection, Wallace perceived it as being an extremely powerful and inexorable force. "The law of Natural Selection or the survival of the fittest is, as its name implies, a rigid law, which acts by the life or death of the individuals submitted to its action,"[20] he wrote in an essay on Darwinism in 1889. In other words, if an animal possessed an inheritable trait that improved its chances in the competition with others, that trait would be favored and enhanced through the generations. A more effective way of digesting food might be a mundane but good example. Conversely, traits that had no particular survival advantage would not be selected for and would not persist and increase through the generations. Wallace turned his sharp criterion on *Homo sapiens,* and found problems.

"I fully accept Mr. Darwin's conclusion as to the essential identity of man's bodily structure with that of the higher mammalia, and his descent from some ancestral form common to man and the anthropoid apes,"[21] he conceded. However, man's intellectual powers and moral sense, among other things, he said, "could not have been developed by variation and natural selection alone, and ... , therefore, some other influence, law, or agency is required to account for them."[22] Darwin was naturally upset by what Wallace called "my little heresy," and he wrote to Wallace in 1869 lamenting, "I hope you have not murdered too completely your own and my child." But Wallace remained steadfast.

His argument was simple and direct. He concluded that if you examine the mental capacity of technologically primitive people —savages he called them, though, if anything, he was less of a racist than his contemporaries—then you find that they are better endowed than they obviously have need for in their simple lives. "Natural selection could only have endowed the savage with a brain a little superior to that of an ape, whereas he actually possesses one but very little inferior to that of the average members of our learned societies."

And what of wit and humor, and mathematical skill, in advanced societies? How could these be the product of natural selection when our forebears could have had no use of them? He listed our peculiarly naked skin as inexplicable by natural selection, our singing voice, our "unnecessarily perfect" hands and feet, and of course our moral sense. "The inference I would draw from this class of phenomena is, that a superior intelligence has guided the development of man in a definite direction, and for a special purpose,"[23] Wallace concluded in 1871, the year in which Darwin published his major statement on human origins, *The Descent of Man.*

It may be that, as Gould has argued, Wallace came to this position by the force of the cold, remorseless logic of the theory of natural selection. And indeed that is how Wallace couches it. But in a long, rambling passage in his 1889 essay on Darwinism you get a distinct glimpse of a man more than a little comforted by what he finds himself able to conclude. "Those who admit my interpretation of the evidence adduced . . . will be relieved from the crushing mental burden imposed upon those who—maintaining that we, in common with the rest of nature, are but products of the blind eternal forces of the universe, and believing also that the time must come when the sun will lose its heat and all life on earth necessarily cease—have to contemplate a not very distant future in which all this glorious earth—which for untold millions of years has been slowly developing forms of life and beauty to culminate at last in man—shall be as if it had never existed; who are compelled to suppose that all the slow growth of our race struggling towards a higher life, all the agony of martyrs, all the groans of victims, all the evil and misery and undeserved suffering of the ages, all the struggles for freedom, all the efforts toward justice, all the aspiration for virtue and the wellbeing of humanity, shall absolutely vanish, and, 'like the baseless fabric of a vision, leave not a wrack behind.' "[24]

Wallace describes the materialistic worldview in which the sun will one day rise no more as a "hopeless and soul-deadening belief." By contrast, his own worldview is full of hope and transcendence. "We, who accept the existence of the spiritual world, can look upon the universe as a grand consistent whole adapted in all its parts to the development of spiritual beings capable of indefinite life and perfectibility. To us, the whole purpose, the only *raison d'être* of the world . . . was the development of the human spirit in association with the human body."

Robert Broom, who, remember, played such an important part in establishing the reality of *Australopithecus* as a part of human ancestry, held a similar but even more extreme view. Not only could he not accept the naturalistic evolution of humanity, but he also could not believe that much of the rest of the complex and beautiful world of animals and plants could have arisen without the intervention of a guiding hand, "a spiritual agency," as he called it. Moreover, he saw the origin of *Homo sapiens* as the ultimate purpose of it all. "Much of evolution looks as if it had been planned to result in man, and in other animals and plants to make the world a suitable place for him to dwell in."[25] And, explicitly influenced by Wallace's writings, Broom finishes on a truly spiritualistic note: "The aim [of evolution] has been the production of human personalities, and the personality is evidently a new spiritual being that will probably survive after the death of the body."

Wallace and Broom therefore accounted for the perceived gap between *Homo sapiens* and the rest of animate nature by employing a "with one bound Jack was free" type of explanation, an explanation that clearly was congenial to their deeply held world views. Others in their profession have employed more scientific explanations, explanations that have nevertheless shifted ground substantially through the past three generations.

There is nothing wrong with explanatory shifts in science: this is the way in which knowledge can advance, moving from one tentative interpretation to the next as new data and new constructs allow. But with human origins, each generation's explanation appears to contain expository themes that go well beyond what might be implied by the new scientific information of the time. "Could it be that, like 'primitive' myths, theories of human evolution reinforce the value-systems of their creators by reflecting historically their image of themselves and of the society in which they live?"[26] asks John Durant, of Oxford University. When he posed this question at a recent annual meeting of the British Association for the Advancement of Science, he was roundly criticized. Hardly surprising, for, like Matt Cartmill, he appeared to be suggesting that what paleoanthropologists do is not very scientific. "Time and again," observes Durant, "ideas about human origins turn out on closer examination to tell us as much about the present as about the past, as much about our own experiences as about those of our remote ancestors."

In fact, insists Cartmill, paleoanthropologists are not necessarily unscientific, because their theories have to be tested against each new piece of evidence, just as happens in other branches of science. "What paleoanthropologists do is more, not less, than scientific," he says. "The mythic dimension is plus, not instead of. The theories still have to resist attempts to prove them false; but they don't *mean* as much without those extensions into the extra-scientific."[27]

Very well, let us examine this progression of ideas.

In the physical realm, any theory of human evolution must explain how it was that an apelike ancestor, equipped with powerful jaws and long, daggerlike canine teeth and able to run at speed on four limbs, became transformed into a slow, bipedal animal whose natural means of defense were at best puny. Add to this the powers of intellect, speech, and morality, upon which we "stand raised as upon a mountain top," as Huxley put it, and one has the complete challenge to evolutionary theory.

Darwin's answer to this was to look at those faculties which appear to make us special—our brains, our bipedality, our use of tools, our sociality—and suggest that, developed little by little, they would give us a competitive edge in the world of brute nature. It was an explanation that made our earliest ancestors already human, albeit to a rudimentary degree. This latter theme has persisted until relatively recently: hominid equals human, and to explain hominid origins is to explain human origins.

For Darwin, the first hominids were brainier than the apes, were more upright than the apes, were more technological and cultural than the apes, and were more social than the apes. In a nutshell, the earliest hominids in Darwin's world were already cultural creatures: they were homunculi. Most of all, they were in competition with the apes and with the rest of animate nature; they were in "the struggle for existence." Darwin even saw an advantage to our ancestors' physical weakness and apparent defenselessness. "An animal . . . which, like the gorilla, could defend itself from all enemies, would perhaps not become social," he suggested.

As befits the inventor of the theory of natural selection, Darwin placed competition at the core of his explanation of human origins, and emphasized its continuing importance. "Man . . . must remain subject to a severe struggle. Otherwise he would sink into indolence, and the more gifted men would not be more successful in the battle of life than the less gifted,"[28] he said. "Darwin's ideas, when applied to human society, were comforting to many other well-to-do Victorians,"[29] observes Matt Cartmill. "Like the idealized moguls of nineteenth-century capitalism, *Homo sapiens* has *earned* mastery of the world by virtue of know-how, shrewdness, and rectitude developed in the 'marketplace' of human competition. Darwinian man is lord of the earth, not because of any God-given stewardship or Romantic affinity to the World Spirit, but for the same good and legitimate reason that the British were rulers of Africa and India."

Darwin's ideas on human origins—in which our "special" attributes were self-explanatory through the incremental advantage of natural selection—persisted into the twentieth century, through the era of Arthur Keith and Henry Fairfield Osborn and on into the

1950s. In this world view, it wasn't so much man's ascendance that puzzled scientists and required explanation, but why the apes had so obviously "failed." The answer was simple: namely, the malign effect of the tropics, "which encouraged indolence in habit and stagnation of effort and growth,"[30] offered Grafton Elliot Smith. "While Man was evolved amid the strife with adverse conditions, the ancestors of the Gorilla and Chimpanzee gave up the struggle for mental supremacy because they were satisfied with their circumstances." There is as much moral disapprobation as scientific explanation in these remarks. *Homo sapiens,* by contrast with the "low" apes, had risen to the highest noble and intellectual plane in the natural world by virtue of his own unrelenting effort.

When, during the 1930s and '40s, the discoveries of australopithecine fossils in South Africa showed that human forebears stood upright and were equipped with small brains as well as small canine teeth, the Darwinian structure began to come apart. Intelligence could not have been an important engine in human evolution if most of the major physical changes in the skeleton had occurred with virtually no expansion in apparent mental capacity. A new explanation was required, and was soon provided. Tool use now emerged as the focus of human advancement, especially tools used as weapons: the era of the killer ape dawned, which was a vastly less flattering self-image than the one of nobility and spirituality enjoyed by Darwin, Keith and their contemporaries.

It was Raymond Dart who set the tone for this new explanatory era, his writings being based on what he judged to be signs of murderous violence in the fossil record. In a landmark paper published in 1953 entitled "The Predatory Transition from Ape to Man," he wrote the following dramatic passage: "The blood-bespattered, slaughter-gutted archives of human history from the earliest Egyptian and Sumerian records to the most recent atrocities of the Second World War accord with early universal cannibalism, with animal and human sacrificial practices, or their substitutes in formalized religions, and with the world-wide scalping, head-hunting, body mutilating and necrophiliac practices of mankind proclaiming this common bloodlust differentiator, this predacious habit, this mark of Cain that separates man dietetically from his anthropoid relatives and allies him rather with the deadliest of carnivores!" As Richard Leakey has commented, "The message of these stirring words is clear: humans are unswervingly brutal, possessed of an innate desire to kill each other."[31]

Dart, being a good biologist, had concluded that the open grass plains of the Transvaal, which is where most of the australopithecine fossils were initially discovered, could not have provided our ancestors with the typical vegetarian diet of the great apes. They must have lived by hunting, he thought, and he saw signs of battered skulls in the fossilized remains of baboons which seemed to support the idea. He also thought he could detect evidence of similarly battered skulls among the australopithecines themselves; hence his speculations about our violent history.

Dart's sanguinary thesis was avidly taken up by playwright Robert Ardrey and, in prose more purple still, transformed into a series of best-selling books that essentially formed a long exposition on the innate depravity of humans and our forebears. "Not in innocence, and not in Asia was mankind born" was how Ardrey opened the first of them, *African Genesis.* Our ancestors hunted for a living, and often turned their murderous talents on their own kind, or so it went. It is what John Durant, of Oxford University, has called the "Beast in Man" hypothesis. "Ardrey," says Durant, "was rewriting the Christian myth of creation in the language of the new biology."[32] Thus was cast the powerful spell of the hunting hypothesis.

Interestingly enough, as Matt Cartmill points out, the essence of the hunting hypothesis had been proposed some thirty years earlier, in various articles published between 1913 and 1921, by two British scientists, Harry Campbell and Garveth Read. But, argues Cartmill, "The world of the 1920s was not prepared to hear about the killer ape. It would take another world war and some additional fossil discoveries to make him and his taste for animal flesh the central issues in paleoanthropological theory."[33] While Campbell and Read's hypothesis was ignored, Dart's words were attended to because they were spoken within a more receptive social context, one in which Sigmund Freud and Konrad Lorenz had laid the foundations of the notion of human depravity from the disparate worlds of psychoanalysis and animal behavior and in which the devastations of the Second World War were still painfully fresh in the collective memory.

Although the more extreme sentiments expressed in Ardrey's writings made professional anthropologists wince somewhat, the core of the hypothesis—that man became man by becoming a hunter—quickly emerged as the new paradigm of human origins. "Hunting is the master behavior pattern of the human species,"[34] said William Laughlin, a University of Connecticut anthropologist, in 1966. He was speaking at a landmark scientific conference of the era, entitled simply "Man the Hunter." The evolution of human hunting, it seemed, could explain everything. "Human hunting is made possible by tools, but it is far more than a technique or even a variety of techniques. It is a way of life,"[35] said Sherwood Washburn and C. S. Lancaster at the same 1966 meeting. "In a very real sense our intellect, interests, emotions, and basic social life—all are evolutionary products of the hunting adaptation."

So, bipedality, intelligence, tool use, culture, and society—all those features which make us human and which had been accounted for by Darwin as the outcome of incremental benefits favored by natural selection—now had a different explanation: hunting. Although different and closer to brute nature than the Darwinian world view, the hunting hypothesis of the 1950s, '60s, and early '70s nevertheless essentially depicted hominid origins as human origins: we are fundamentally human from the start. "*Australopithecus,*" said a prominent anthropologist of the time, "was our kind of animal."

Given this equation of hominid with human, it is perhaps not surprising that anthropologists should be more than a little sensitive about the nature of the behavior and relationships of even the earliest of the hominid line: simply put, self-image was at stake. For Arthur Keith and his colleagues, an important defensive measure had been to see an essentially modern form of man disappearing deep into prehistory. By this means a comfortable distance was interposed between humanity and the beast. For anthropologists of the 1960s and early '70s, the same effect was achieved by pushing hominid origins back as far as they could be stretched, thus keeping the ape in us at a nonthreatening distance.

From the mid 1970s on, the hunting hypothesis and all that it implied began to fall apart, for a number of different reasons. First of all, with new and spectacular discoveries made in East Africa, it began to be clear that the first stone tools in the archeological record do not begin to appear until at least a million years after the earliest hominids had already evolved a fully bipedal gait. In the absence of stone tools as weapons and butchery implements at the beginning of the human line, the argument for hunting as the driving force behind the origin of bipedalism simply vanished. As a result of a subsequent reexamination of the archeological evidence, paleoanthropologists now suspect that fully developed hunting of the sort that so fired the collective imagination a decade ago was adopted only very recently in human history. It may be that our forebears were opportunistic scavengers, not hunters, for most of their career—an idea that many find most unflattering to our self-image.

A second major development has been the growing realization of the true implications of a many-branched hominid family tree. Some of the australopithecine species were anatomically extremely robust, with massive cheek teeth and heavily muscled jaws, which is in stark contrast with the more delicately boned early *Homo* species. When examples of both these types of hominid were found more or less literally cheek by jowl on the eastern shore of Lake Turkana in 1973, anthropologists finally began to think in terms of two very different kinds of animals, two very different kinds of ecological niches. No longer was it acceptable to talk of "the" hominid adaptation, because there clearly were several. And as the earliest members of the *Homo* line apparently arose at least a million years later than the earliest known hominid, *Australopithecus afarensis*, it was no longer acceptable to equate hominid origins with human origins. Whatever it was that eventually made us human was apparently quite unconnected with what had initially caused the first hominids to adopt upright walking and to lose their daggerlike canine teeth. Adducing human attributes—such as intelligence and culture—to explain the origin of these first hominid adaptations was therefore no longer relevant. By the same token, our self-image is less threatened by the distinctly primitive and apelike nature of our earliest ancestors.

From the mid 1970s onward, the hunting hypothesis was also attacked from theoretical standpoints. One, developed by the late Glynn Isaac and promulgated by Richard Leakey in several popular books, emphasized sharing and cooperation as the key behavioral ingredients in hominid origins and human success. Owen Lovejoy, meanwhile, suggested that demographic and nutritional demands spurred the development of bipedalism and monogamous bonding between males and females. As a counterpoint to the male-oriented hunting hypothesis, Adrienne Zihlman and Nancy Tanner suggested that the mother/infant bond and food-sharing among mature females were at the core of hominid origins.

Whatever the relative merits of these various proposals—and it is not easy to test all of them in the record—each has the clear intention of replacing a distinctly aggressive image of human origins with a distinctly peaceable one. "But why are people trying so hard to do that?"[36] asks Matt Cartmill. "The striking thing about these theories is that they go so far beyond the available evidence in an effort to show that hunting was *not* important in early hominid evolution—just as the killer ape theorists did to prove that it was crucial." Why? What lies behind it? "When people turn indignantly from one sort of speculation to embrace another, there are usually good, nonscientific reasons for it," Cartmill observes.

These reasons might include an attempt to turn away from the pessimistic view that humans are bound by their very nature to annihilate themselves through the agency of nuclear war. Or to reject the idea that through our heritage, we are innately programmed to behave in any particular manner at all, and especially in an undesirable manner. But in the long run it is of little account what these reasons are, because they are reasons of the moment. They are, as John Durant says, "a direct response to contemporary social experience."[37] These peaceable theories of human origins, like the beast-in-man idea, become "a mirror which reflected back only those aspects of human experience which its authors wanted to see. . . . This is precisely what we would expect of a scientific myth."

Most scientists wince when the word "myth" is attached to what they see as a pursuit of the truth: science, remember, is supposed to be objective, and truth is spelled with a capital T. "A myth, says my dictionary, is a real or fictional story that embodies the cultural ideals of a people or expresses deep, commonly felt emotions," observes Cartmill. "By this definition, myths are generally good things—and the origin stories that paleoanthropologists tell are necessarily myths. They are myths whether they are true or not, because they embody a fundamental cultural theme: they define and explain the crucial difference between human beings and beasts."[38]

The Truth about man's place in nature is therefore to be sought in four quite separate dimensions. In the first three levels—of time, form, and behavior—there is scientific evidence, from fossils, stone tools, comparative anatomy and behavior, and molecular biology. Using this evidence, it may one day be possible accurately to draw lines back through time, connecting ourselves with our forebears, their forebears with theirs, and so on until a detailed evolutionary tree traces the link between humanity and brute nature.

Exactly where brute nature ends, however, and humanity begins is not a question for molecular or comparative biology. It is a question of the fourth dimension: a question of self-image. Here there are no lines accurately to be drawn, no hypotheses to be tested, for humanity's view of itself is constantly shifting, depending on the experience of the moment.

Paleoanthropology has, and always has had, as its major goal the search for man's place in nature. The science shares with all historical sciences the limitations of trying to reconstruct events that occurred just once: there are no experiments to be done that can confirm or deny the major themes that are sought. It also shares with all sciences the truism that science is an activity done by people, and is therefore subject to the unavoidably personal and erratic nature of intellectual progress. But paleoanthropology alone among all the sciences operates within the fourth dimension, with humanity's self-image invisibly but constantly influencing the profession's ethos.

As Matt Cartmill said, "All sciences are odd in some way, but paleoanthropology is one of the oddest."[39] For this reason, there will always be bones of contention.

Afterword

The decade since the publication of the first edition of *Bones of Contention* has been one of the more productive periods in the history of paleoanthropology. For instance, teamwork analysis of the virtually complete, 1.6-million-year-old skeleton of a young, early *Homo* individual that was found in northern Kenya in 1985 has yielded unprecedented insights into the anatomy and behavior of early humans. In addition, the first early human fossil has been found west of the Rift Valley, in Chad, and named a new species of *Australopithecus*. Two new species of early humans also have been named, based on fossil discoveries in Ethiopia and Kenya. These discoveries place the known fossil evidence at 4.4 million years old, a million years older than prior findings. And the earliest history of the genus *Homo* is now said by some to represent several species, not just one, as had been thought previously.

This proliferation of species may seem reminiscent of the "splitting" proclivity prevalent among anthropologists three or more decades back, when every new fossil discovery was labeled a new species (as described in chapters 5 and 7). Indeed, there is as yet no consensus about the proposed new species and their evolutionary relationships, and there is continued resistance to bushy rather than ladderlike evolutionary trees in our history. In fact, the recognition of more, rather than fewer, species as branches of the human family tree is serving to make anthropology more solidly based in biology, because this is the pattern common to the evolution of most mammalian groups. That paleoanthropology is becoming more scientific is a healthy sign because, as described in earlier chapters, a pervasive practice in paleoanthropology has been to view human evolution as somehow different from the evolution of other animals. This view derives from the sentiment that humans are special. We are, of course, in many ways, but it has been all too easy to impose the perspective of what it is like to be *modern* humans on the interpretation of our evolutionary history. Later, I will propose that this tendency continues still in some quarters.

Perhaps the most lively, and controversial, realm of paleoanthropology during the past decade, however, concerns the origin of modern humans: that is, when and where did people like us evolve? There have been hundreds of research papers and scores of books and conferences devoted to the topic in the past decade—a measure of its importance to anthropologists. Coverage in the popular media—in newspaper and magazine articles and in books—has also been extensive, which suggests that the topic is especially relevant, or at least fascinating, to a more general audience. One reason, undoubtedly, is that differences of opinion among scholars have been sharp, and sharply expressed. This debate certainly qualifies as a controversy in the context of this book, and controversies make for "good copy" in the popular media. But another reason is that, because we humans are such great egotists, few things are more interesting to us than the origin of ourselves. Not surprisingly, therefore, the research into the origin of modern humans is fertile ground for the intrusion of subjectivity.

The history of the debate began with the first discovery and interpretation of Neanderthal fossils in the middle of the nineteenth century. Specifically, the question was this: were the Neanderthals our direct ancestors, or extinct cousins? For more than a century, the debate focused on fossil evidence, with specimens from Europe, Africa, and Asia forming the objects of analysis. Opinion swung back and forth throughout that time. Then, in the 1980s, evidence from molecular biology joined the debate. An already controversial subject became yet more so, with the molecular evidence apparently exclusively supporting one of the competing theories based on fossils. My aim here is not to detail the history of this increased debate in its entirety, as this would demand another book. Instead, I will look at some of the language and imagery that attended the impact of the molecular evidence, with some observations of how the science affected the language, and how the language affected the science. Some history and theoretical context is necessary, however.

When the molecular evidence entered the debating arena a decade ago, there were two principal, distinctly opposing theories of the origin of modern humans. One, called the multiregional evolution hypothesis, argued that modern humans evolved from populations of *Homo erectus* that had expanded beyond Africa almost 2 million years ago and had become established in many regions of the Old World. Under this hypothesis, modern humans are the product of a near-simultaneous evolution over a wide geographical area. The second hypothesis, called the Out of Africa hypothesis, argued instead that modern humans evolved fairly recently in a single population in Africa, and that descendants of this population had expanded into the rest of the Old World, replacing existing populations of archaic *sapiens,* such as Neanderthals, with little or no interbreeding.

The multiregional hypothesis was the first comprehensive theory of the origin of modern humans, and its history goes back more than 50 years to the German anatomist Franz Weidenreich. (There are even deeper roots, but Weidenreich is a good starting point.) Weidenreich envisaged parallel evolutionary lineages in various regions of the Old World leading through separate Neanderthal-like stages (archaic *sapiens*) to the geographical variants of modern humans. Weidenreich's proposal came to be known as the "candelabra model" of modern human origins: drawn schematically, the long regional ancestries look like an array of candles.

Weidenreich was aware that, because he had suggested that each modern geographical population traces its origins back through Neanderthal-like and *Homo erectus* precursors, modern races might be thought of as having separate origins, even as being separate species. In 1949 he explicitly stated that this was not his view. Nevertheless, in 1962, the University of Pennsylvania anthropologist Carleton Coon came close to proposing what Weidenreich had warned against. Coon argued not only that racial differences were ancient, but also that some races had achieved sapienshood earlier than others. The notion that extant racial groups have been genetically separate for at least a million years and that some evolved relatively recently lent itself readily to the inference of deep differences between the races.

Despite Weidenreich's efforts, the so-called unilinear point of view was slow to take hold. Eventually, however, a confluence of events occurring from the 1940s through the 1960s led to the development of several competing hypotheses, of which the unilinear hypothesis was one. Another included the proposition of the evolution of modern forms earlier than Neanderthal, which therefore left Neanderthals as an extinct side branch. Yet another suggested that the "classic Neanderthals" of western Europe were the evolutionary product of less extreme forms in eastern Europe and the Middle East, which also gave rise to modern humans.

The emergence of the unilinear hypothesis as the most prominent among its competitors came about in the 1960s, principally through the efforts of Loring Brace of the University of Michigan. In a 1964 paper, "The Fate of the 'Classic' Neanderthals,"[1] Brace argued persuasively that Neanderthal anatomy had been mistakenly interpreted as extreme, and that it could be seen as ancestral to that of later European modern people. By the late 1960s, therefore, the Neanderthals had been restored to what was, in many people's eyes, their rightful place as direct ancestors of modern humans. Fossils that had been discovered in Europe, Africa, and Asia during the first half of the century were now interpreted by Brace and his supporters within the unilinear theory as evidence of evolution toward *Homo sapiens* in many different parts of the Old World.

Brace argued that the adaptive environment of humans was "a cultural niche." In other words, the same kinds of anatomical modernization were going on in populations throughout the Old World as a result of the adoption of a more advanced form of technology. With a common technological context, the evolutionary context of all humans was essentially the same, and must therefore proceed in the same direction, because the natural environment had become less important. According to Brace's "single species hypothesis," only one species of human existed at any given period in evolution—the ultimate expression of the unilinear pattern. Milford Wolpoff, also at Michigan, joined Brace as a vigorous spokesman for the hypothesis. However, in the mid-1970s, the discovery of the coexistence at Koobi Fora, northern Kenya, of a small-brained, highly robust individual (KNM-ER 406, *Australopi-*

thecus boisei) and a large-brained, nonrobust individual (KNM-ER 3733, *Home erectus*) demolished the single species hypothesis, at least for the period of human prehistory close to 2 million years ago.

Wolpoff, now a major protagonist in the current debate on the origin of modern humans, nevertheless insists that the unilinear hypothesis holds for the later stages of human prehistory. The multiregional evolution hypothesis offers an explanation not only of the origin of *Homo sapiens* but also the existence of anatomical diversity in modern geographical populations. This diversity is said to be the result of the evolution of distinctive traits (through adaptation and genetic drift) in different geographical regions that became established in early populations of *Homo erectus* and persisted through to modern people. This persistence is known as regional continuity.

In its original formulation by Weidenreich, the multiregional hypothesis posited limited gene flow (mating) between different geographical populations. Wolpoff, joined by Alan Thorne of the Australian National University, Canberra, argues that gene flow between populations has been important. The hypothesis therefore views the *erectus*-to-*sapiens* transformation as a balance between the maintenance of distinctive regional traits in anatomy through partial isolation of populations that persisted through very long periods of time, and the maintenance of a genetically coherent network of populations throughout the Old World through significant gene flow, or mating between populations.

The Out of Africa, or recent, single origin hypothesis, has a shorter history, going back to the ideas of Louis Leakey in the 1960s. He considered the Early and Middle Pleistocene humans of Africa to be better candidates for modern human ancestry than the *Homo erectus* fossils of Asia; the latter, he said, were an evolutionary dead end. The most extreme version of the modern form of the Out of Africa hypothesis, which includes the substantial replacement of archaic populations by invading modern humans, is most closely associated with Christopher Stringer of the Natural History Museum, London. This model accepts some interbreeding between archaic and early anatomically modern populations, but sees the long term effects of such interbreeding as minor; it views the establishment of regional anatomical traits in today's geographic populations as the result of adaptation and genetic drift in local populations during the last 100,000 years. Other current hypotheses encompass a single (probably African) origin, but differ from the above version in the degree of genetic hybridization between archaic and early anatomically modern populations.

An interesting consequence for taxonomy has flowed from the multiregionalists' view of the mode of evolution of modern humans. They see the evolutionary transformation as continuous change within a genetically coherent lineage. This being the case, there is no clear break between *Homo erectus* and *Homo sapiens*, merely an evolutionary continuum. As a result, say the multiregionalists, there is no valid reason to distinguish between species, but instead the lineage should be regarded as a single species, that of *Homo sapiens*. Not surprisingly, many scholars question the biological reality of lumping together, say, the robust *Homo erectus* people of Java and the gracile people of today's world into the same species.

In January 1987, Allan Wilson and two colleagues from the University of California, Berkeley, published a paper in the journal *Nature* titled "Mitochondrial DNA and Human Evolution." It was, it might be said, the shot that was heard around the anthropological world, for it ignited into a conflagration the battle that was already smoldering between supporters of the multiregional evolution and Out of Africa hypotheses. In the paper, Wilson and his colleagues wrote that "[the] transformation of archaic to anatomically modern forms of *Homo sapiens* occurred first in Africa, about 100,000 to 140,000 years ago, and that all present-day humans are descended from that African population."[2] Wilson and his colleagues also noted that there was no evidence of a genetic mixing (or mating) between the incoming modern humans and established populations of archaic humans, in the form of ancient lineages of mitochondrial DNA in today's population. In other words, there apparently was complete replacement of existing archaic populations by incoming modern humans. The Berkeley work thus offered strong support for the Out of Africa hypothesis and none for the multiregional evolution hypothesis.

Mitochondria are the energy-producing organelles present in all cells in the body. They contain circular molecules of DNA, which in humans measure 16,569 base-pairs in length. Mitochondrial DNA possesses several properties that are useful in reconstructing recent evolutionary histories of modern populations. First, it is inherited strictly through the maternal lineage, making reconstruction of evolutionary patterns more straightforward than nuclear DNA, which recombines from both parents at each generation, thus scrambling the DNA to some degree. Second, mitochondrial DNA accumulates mutations at about ten times the rate of nuclear DNA and therefore offers a "fast-ticking" molecular clock that can record relatively recent events. In terms of evolutionary time, events within the last several hundred thousand years are relatively recent.

By analyzing about 9 percent of the DNA sequence in 147 individuals from around the world (using a technique known as restriction fragment length polymorphism), Wilson and his colleagues found that (1) as previously stated, all the DNA types were of recent origin; (2) linking the types of DNA in a genealogical tree produced two groups, one containing representatives of all populations and a second containing representatives only of African populations; and (3) the amount of genetic variation in the DNA types was greater in Africans than in any other population. These findings were the foundation of their strongly stated conclusion.

Tracing the genealogical tree of DNA types effectively leads back to the DNA of a single female who lived many generations ago. Imagine a population of, say, 5,000 mating pairs, each with a different family name. Now imagine that as time passes the population remains stable; each couple produces only two offspring. In each generation, on average, one quarter of the couples will have two boys, one half will have a boy and a girl, and one quarter two girls. In the first generation, therefore, one quarter of the family names will be lost. As each generation passes, more losses will occur, but at a slower rate. After about 10,000 generations (equal to twice the number of original females) only one name will remain. The same pattern holds for the loss of mitochondrial DNA types, except the transmission is through the female line. For this reason, Wilson often described the single female from whom we all derive our mitochondrial DNA in the modern world as "one lucky mother," because the dynamic of DNA loss is stochastic.

This explanation is a long, but unavoidable, way of coming to an aspect of the language of Wilson's argument that unfortunately muddied the debate. A year before the publication of the *Nature* paper, a reporter for the *San Francisco Chronicle* wrote a front-page story about the research of Wilson and his colleagues, titled, "The Mother of Us All—A Scientist's Story." Charles Petit, the reporter, described the Berkeley team's conclusions and noted that "the dramatic, controversial claim of a fairly recent African Eve as the very very great-grandmother of all humans is sure to stir up an old debate in paleoanthropology."[3] He was referring to the debate over the origin of modern humans, and he was right in his prediction. However, by using the name "Eve" he also added an extra dimension to the debate.

Now, had Petit referred to that single female as the "mitochondrial DNA lineage coalescent point," which is the technically correct term, there is little doubt that the story would not have attracted the attention it did. But invoking the name "Eve" led writers to describe her in ways that the evidence did not imply, as we'll see below. Most important, because the terminology conjures up the notion of Adam and Eve in the Garden of Eden, it influenced the scientific argument; that is, the way scientists, not just newspaper reporters, wrote and spoke of the issue.

No one ever said there was just one breeding pair of individuals in the origin of modern humans, as in the Garden of Eden story, but the seductive notion of Eve implied that there were very few individuals—a few tens, perhaps, or a few hundred at most. When populations fall to low numbers like these, the result is described as a population bottleneck. The idea of an extreme bottleneck, as this low population number certainly would suggest, therefore became identified with the Mitochondrial Eve hypothesis, as it came to be called. Wilson did toy with the idea of a bottleneck at the origin of modern humans, but only briefly. Very soon he said Eve had been just one of perhaps 5,000 females in a population twice that size.

Nevertheless, critics of the Eve hypothesis were quick to invoke genetic evidence that seemed to argue against the extreme bottleneck in recent human history as evidence against the validity of the hypothesis. The argument continues today, as in a paper in the journal *Science* by the University of California, Irvine, geneticist Francisco Ayala, titled "The Myth of Eve." Although he concedes that "molecular evolution data favor the African origin of modern humans," he goes on to say that "the weight of the evidence is against a population bottleneck before their emergence."[4] Ayala was arguing against a population of far less than 5,000 breeding females, even though the Eve hypothesis does not demand this condition.

In any case, Eve quickly caught the media's imagination and even received the ultimate public accolade, becoming the subject of a cartoon in the *New Yorker*. Eve also made a cover story in *Newsweek*, a year after the work was published in the scientific literature. The *Newsweek* text acknowledges that Eve was not the only female living 150,000 years ago, but it suggests that she "was the most fruitful" and that she left behind "resilient genes" that are carried by all of humankind.[5] In retrospect it might appear that way, but in fact, as I've explained, the survival of the mitochondrial DNA lineage has to do with pure chance, not greater fertility or superior genes. But we'd expect Eve, Mother of us all, to be fruitful and strong, wouldn't we?

For the next several years, anthropologists struggled with this new genetic evidence, particularly the multiregionalists whose hypothesis it undermined. Here, it seemed, was a coterie of molecular biologists, arrogant as ever, strutting onto the scene, effectively saying, "You want an answer to your question? Here's an answer from DNA. Your fossils just aren't up to it!" A passage from an article in *Discover* in 1990, written satirically, was meant to explain anthropologists' negative reaction to Eve:

> "Her name was Eve, and she was trouble from the start. Sure, the public loved her: she was brash, sexy, and surprising, with a body of data you could reach out and grasp and implications that just wouldn't quit. She made the cover of *Newsweek*. She even got on Johnny Carson. Not too shabby for a human-origins hypothesis born in a biochemist's beaker. But for paleoanthropologists, the hard-boiled types who earn their living making dead men talk, Eve spelled poison forward and backward. She was an interloper, a biochemical bauble, a dolled-up set of assumptions masquerading as a breakthrough. To them, the only good Eve was a refuted Eve."[6]

And refutation is what appeared to have happened at a meeting of the American Association for the Advancement of Science in New Orleans in February 1990. A group of anthropologists organized a symposium that purported to offer a balanced view of the fossil evidence, stating that the fossil data were "the only direct evidence within the power of refutation." David Frayer of the University of Kansas summarized the group's deep negative reaction to the molecular evidence when he stated, "Fossils are the real evidence." The group concluded that the fossil evidence is unequivocal: "it proves that the Eve hypothesis *must* be wrong." The symposium was widely reported, and a quote from *The Times* of London gives a taste of the tenor of those reports: "A group of specialists on fossils said that studies of skulls and other remains of humanlike creatures in Asia and Europe showed that the Garden of Eden theory must be wrong."

The anthropologists had prepared careful summaries of their arguments and handed them out as a single document to the journalists at the meeting. The document was titled: "Eve: the fossils say No." What the document didn't explain was that every one of the speakers was a proponent of the multiregional evolution hypothesis and so had a vested interest in proving the Eve theory wrong. A different group of anthropologists, similarly selected by, say, Christopher Stringer, would have reached a different conclusion: "Eve: the fossils say Yes."

Milford Wolpoff was the symposium's organizer, and later he told a reporter that the event was "a sales pitch," not the objective assessment of all the fossil evidence as the program had implied. "We planned the whole thing, rehearsed it, worked over the exact phrasing," he explained. "We felt we had to do this, because we were becoming victims of the complexity of our ideas. . . . We think the fossils absolutely support our position."[7]

There's nothing wrong with careful preparation, of course. That makes for clear communication, particularly to reporters who may not be familiar with the details of the debate. But a sales pitch that may be inferred by the audience to be a balanced scientific presentation of all the fossil evidence is something different. Nevertheless, it had the desired impact on the newspaper stories.

Wolpoff used some colorful language in his aforementioned discussion with the reporter. "Our hypothesis of regional continuity may not be as sensational as this idea of killer Africans sweeping out across Europe and Asia, over-running everybody," he is reported as saying.[8] The language was apparently meant to cast the Out of Africa hypothesis in a bad light. "There is no way one human population could replace everybody else except through violence," continued Wolpoff. "The people advocating replacement have to come to terms with what they are saying. I'm just glad it's them and not me."[9] Parallels with the near genocide of native peoples in the Americas and Australia in recent history are obvious. Christopher Stringer told the same reporter, "It simply isn't true. There are plenty of ways for a population to become extinct without being murdered by another population."[10]

In fact, Ezra Zubrow of the State University of New York, Buffalo, has developed demographic models that show that violence is *not* necessary for rapid population replacement. In modeling competition for resources between populations of, say, Neanderthals and modern humans, he shows that "a small demographic advantage in the neighborhood of two percent mortality [of moderns over Neanderthals] will result in the rapid extinction of the Neanderthals."[11] By "rapid," Zubrow means within a single millennium, which is consistent with the fossil record.

Meanwhile, Wilson and his colleagues had been facing criticisms of their initial study. Opponents pointed to the fact that they used only a limited amount of genetic information from the mitochondrial genome, and that the sample of "Africans" were African Americans, not native Africans. Moreover, one worker had reanalyzed their data and found interpretations that did not include an African origin as a necessary conclusion. Wilson and his colleagues responded by obtaining native African genetic material and sequencing a section of the mitochondrial genome. They subjected the data to more rigorous statistical analysis. They concluded in September 1991, "Our study provides the strongest support yet for the placement of our common mitochondrial DNA ancestor in Africa some 200,000 years ago."[12] The result was said by the authors to be statistically significant, a powerful claim in biology and particularly in paleoanthropology.

Within six months, the *New York Times* was proclaiming, "Critics Batter Proof of an Ancient African Eve." The text said, "The serpent of uncertainty has slithered into the garden. Its bite has undermined a critical statistical foundation for the Eve hypothesis."[13] Wilford, one of the best-informed and most thoughtful of newspaper science writers, could not resist developing the Garden of Eden imagery, and rightly so. Continuing the biblical imagery, an article in *The Chronicle of Higher Education* stated that "Black Eve had fallen from grace."[14]

The new problem that the Berkeley team's analysis faced was the statistical nightmare of inferring an evolutionary tree from the vast amount of genetic data that they had fed into the system. Trees are reconstructed using the parsimony approach, a common practice in molecular phylogeny. Simply put, the technique looks at the genetic variation among modern populations and tries to build trees with the fewest number of genetic changes that link these variants. Wilson and his colleagues believed that the simplest trees were rooted in Africa. In fact, with the volume of genetic data to hand, and the shallow differences among individuals, there are millions of possible trees, so many that it is impossible with our current computation power to test them all. Short cuts are necessary, and, unfortunately for Wilson and his colleagues, they made what was an extremely complex picture look simple. (Wilson died about this time, leaving his colleagues, particularly Mark Stoneking, now at Pennsylvania State University, to carry the torch.)

By this time a cottage industry had grown up dedicated to unraveling the mitochondrial DNA story, with some workers gathering and analyzing more data and others reanalyzing existing data. Prompted by the scrutiny of others, Stoneking and his new colleagues conceded in a short item in *Science* in February 1992 that the conclusions about an African Eve could no longer be said to be statistically significant, but he held that the original conclusion was likely to be true. Wolpoff demurred: "It's over for Eve," he said at a conference not long after Stoneking's *Science* paper.[15] A more dispassionate assessment was that by David Hillis of the University of Texas: "The data are ambiguous. They don't argue that there wasn't an African origin, and they don't argue that there was one. It's like saying you can't solve a mystery after reading one page of the book."[16]

This setback for Mitochondrial Eve forced molecular biologists to accept that unravelling modern human origins with genetic data was not as straightforward as they had believed. Mitochondrial DNA was clearly not going to solve the issue, not least because it is effectively one gene, even though it is a long piece of DNA containing more than a dozen genes. The reason is that the mitochondrial genome is inherited intact, like a single gene. As Stoneking noted in his *Science* note, "DNA sequence data from multiple nuclear genes, in combination with mtDNA data, likely will be needed to overcome the effect of single gene phylogenies."[17]

Since that time, several laboratories around the world have been producing detailed genetic data on half a dozen nuclear DNA sequences, including some from the Y (male) chromosome. Because the Y chromosome is present only in males, it is in effect the male equivalent of the mitochondrial DNA genome, yielding information on male lineages. This, and the great majority of the other nuclear DNA examined, is consistent with a recent origin for modern humans and provides strong support for Africa as the region of origin.[18] So, although the original use of mitochondrial DNA information did not yield unequivocal conclusions, the conclusions it did lead to have been overwhelmingly corroborated. Proponents of the Out of Africa model (based on fossil evidence) can therefore gain comfort from these conclusions, while multiregionalists are left to explain the inconsistency with their hypothesis.

In an ironic twist, a relatively recent technique for analyzing mitochondrial DNA returns to the issue of a population bottleneck in human prehistory. The technique, known as mismatch distribution, essentially compares sequence differences between pairs of individuals in modern populations. One of the striking features of mitochondrial DNA in modern humans is the small degree of variation among populations, compared, for instance, with that in our closest relatives, the African great apes. For instance, the degree of genetic difference between, say, an Eskimo in Alaska and a Pygmy in Africa is one-tenth as great as that between two gorillas in the same population in a single section of forest in Uganda. According to mismatch distribution analysis, the reason for this low genetic variability is that human populations were extremely small until about 60,000 years ago, when they expanded dramatically. How does this inform the debate over the competing hypotheses for the origin of modern humans?

As with the other molecular evidence, this new analysis offers support for the Out of Africa model and little comfort for the multiregional evolution model. For instance, the human population is estimated to have been only a few thousand some 100,000 years ago. This poses a severe practical problem for multiregionalists. "It is difficult to imagine that a population this small could have populated all of Europe, Africa, and Asia," as required by the multiregional evolution hypothesis, conclude Alan Rogers and Lynn Jorde of the University of Utah.[19] "We conclude that all available genetic evidence is consistent with the proposition that the major human populations separated from a small initial population roughly 100,000 years ago and that most of these separate populations experienced a bottleneck . . . several tens of thousands of years later."[20] The size of the population envisaged, a few thousand, is compatible with genetic evidence that rules out *extreme* bottlenecks in human prehistory.

Charles Petit's evocation of Eve therefore might not have been so far wrong after all, but it has taken a decade to work through mountains of data and esoteric methods of analysis to take anthropologists back to the Garden of Eden. They bit the apple of the Tree of Knowledge; they found themselves enmeshed in a situation far more complicated than they had expected.

The story is effectively over, as the majority of anthropologists now support some form of recent origin hypothesis. Wolpoff and many colleagues, however, still adhere to the multiregional evolution hypothesis. We are left with the position so often reached earlier in the book: two groups of anthropologists, faced with the same fossil evidence, come to diametrically opposed conclusions. Multiregionalists interpret the fossil evidence as demonstrating that anatomical features present a million or more years ago in humans in different parts of the world pass in an uninterrupted sequence through to modern populations. Proponents of the Out of Africa hypothesis see the same anatomy as indicating that features of modern populations originated relatively recently, first in Africa and then spreading into the rest of the Old World. One group or the other is wrong, and the genetic evidence implies strongly that it is the multiregionalists.

It has to be said that in no other mammalian group is there a known evolutionary pattern that is equivalent to that proposed for the multiregional evolution of modern humans. Humans, the hypothesis implies, are special and different. In separate populations throughout much of the Old World, an identical trajectory of genetic change is said to have taken place over a period of at least a million years, and this includes a significant increase in the size of the brain. Our supposed specialness, therefore, includes this inexorable trend to increased intelligence. The quality of high intelligence itself surely sets us apart from other animals, but why should the evolutionary mechanism of its origin also separate us from other animals?

Perhaps the multiregional evolution hypothesis should be seen as species-centric, in the same way that anthropologists three decades ago were species-centric when, as discussed in chapters 5 and 6, they were loath to accept that we shared a common ancestor with apes a mere 5 million years ago, not 30 million years as was then held to be the case? In a recent popular book, Christopher Stringer suggests this is so. The multiregionalists' position, he suggests, is "a last vestige of this urge to self-importance."[21] But Stringer has much invested in the Out of Africa hypothesis being correct, so he would be expected to take such a view, wouldn't he?

And the history of the debate over the origin of modern humans has swung back and forth between models like the multiregional evolution and Out of Africa hypotheses, so perhaps in the future it can be expected to swing back in the multiregionalists' favor? Nevertheless, there is just one true history of modern humans—the one that happened over the past million or so years—and the task is to move toward it and recognize it when is is reached. Scientific theories are tentative constructs, it is true, but they are not forever oscillating like a pendulum.

Notes

CHAPTER 1

1. CBS Inc. 1981. All rights reserved. Originally broadcast in May 1981 over the CBS Television Network as part of the *Universe* program series.
2. Both Leakey and Johanson have published popular books in collaboration with science writers. Leakey's, which include *Origins* (1977), *People of the Lake* (1979), and *The Making of Mankind* (1981), were written with Roger Lewin, and Johanson's, *Lucy*, with Maitland Edey.
3. Interview with the author, Nairobi, 21 January 1985.
4. As 1.
5. As 3.
6. As 1.
7. As 1.
8. Interview with the author, Berkeley, California, 19 November 1985.
9. Sir Peter Medawar, "Induction and Intuition in Scientific Thought," reprinted in *Pluto's Republic*, Oxford University Press, 1984, page 78.
10. "Four Million Years of Humanity," lecture at the American Museum of Natural History, New York, 9 April 1984.
11. As 3.
12. *The Roots of Mankind*, published by George Allen & Unwin, 1971, page 139.
13. *Essays on the Evolution of Man*, published by Oxford University Press, 1924, page 55.
14. *Smithsonian Report* for 1927, pp. 417–32.
15. "Four Legs Good, Two Legs Bad," *Natural History*, November 1983, page 65.
16. *Smithsonian Report* for 1928, page 416.
17. As 10.
18. Interview with the author, Berkeley, 19 November 1985.

19. As 10.
20. As 8.
21. Interview with the author, Berkeley, 2 October 1984.
22. Interview with the author, London, 11 June 1985.
23. *Apes, Men and Morons*, published by Putnam, 1937, page 112.
24. "Reflections on Human Paleontology," in *A History of Physical Anthropology: 1930–1980*, published by Academic Press, 1982, page 231.
25. *Man-Apes or Ape-Men*, published by Holt, Rinehart and Winston, 1967, page 9.
26. "Choose Your Ancestors," lecture at the California Institute of Technology, Pasadena, September 1974.
27. "Myths and Methods in Anatomy," *Journal of the Royal College of Surgeons of Edinburgh*, vol. 11, no. 2, pp. 87–114 (1966), page 91.
28. As 10.
29. As 27, page 113.
30. As 10.

CHAPTER 2
1. Interview with the author, Boston, 22 January 1986.
2. "Human Evolution as Narrative," in *American Scientist*, vol. 72, pp. 262–68 (1984), page 265.
3. *A New Theory of Human Evolution*, published by the Philosophical Library, New York, 1949, page 161.
4. *Meet Your Ancestors*, published by John Long, Ltd., New York, page 10.
5. *Essays on the Evolution of Man*, published by Oxford University Press, 1924, page 79.
6. "Recent Discoveries Relating to the Origin and Antiquity of Man," *Science*, vol. 65, pp. 481–88 (1927), page 482.
7. *Man Rises to Parnassus*, published by Princeton University Press, 1927, page 164.
8. As 7, page 79.
9. "The Trend of Evolution," in *The Evolution of Man*, published by Yale University Press, 1922, pp. 152–84.
10. As 5, page 40.
11. "Four Legs Good, Two Legs Bad," *Natural History*, November 1983, pp. 65–78. page 68.
12. "Aspects of Human Evolution," in *Evolution from Molecules to Man*, edited by D. S. Bendall, published by Cambridge University Press, 1983, page 515.
13. Review of Landau ms. for *American Scientist*.
14. As 12, page 515.
15. As 1.
16. Interview with the author, Berkeley, 3 October 1984.
17. Letter, Washburn to Landau, 29 April 1981.
18. Letter, Washburn to Landau, 14 May 1981.
19. As 16.
20. As 1.
21. As 2, page 262.
22. As 1.

23. Ms. of lecture, "Paradise Lost," given to symposium "The Rhetoric of the Human Sciences," University of Iowa, 28–31 March 1984, page 2.
24. *Scientific Monthly*, vol. 39, page 486 (1934).
25. As 23, page 2.
26. "The Locomotor Behavior of *Australopithecus afarensis*," in *American Journal of Physical Anthropology*, vol. 60. pp. 279–317 (1983).
27. "Human Evolution: The View from Saturn," in *The Search for Extraterrestrial Life: Recent Developments*, IAU, 1985, pp. 213–21.
28. *The Myths of Human Evolution*, published by Columbia University Press, 1982, page 2.
29. As 5, page 77.
30. *Bulletin of the New York Academy of Medicine*, III, pp. 513–21 (1927).
31. As 5, page 68.
32. *The Coming of Man: Was It Accident or Design?* published by H. F. & B. Witherby, London, 1933, page 10.
33. As 32, page 220.
34. As 32, page 218.
35. "The Baron in the Trees," a presentation to conference on "Variability and Human Evolution," Rome, 24–26 November 1983, ms., page 11.
36. "The Dawn Man of Piltdown, Sussex," in *Natural History*, 21, p. 577 (1921), page 578.
37. As 35, page 4.
38. "Current Argument on Early Man," in *Major Trends in Evolution*, edited by Lars-Konig Konigson, published by Pergamon Press, 1980, pp. 261–85, page 262.
39. As 38, page 267.
40. As 38, page 262.
41. As 11, page 77.
42. As 38, page 262.
43. "*Australopithecus africanus:* The Man-Ape of South Africa," in *Nature*, vol. 115, page 196 (1925).
44. As 35, page 9.
45. "A Systematic Assessment of African Hominids," in *Science*, vol. 203, pp. 322–33 (1979).
46. As 35, page 10.
47. "The Myth of Human Evolution," in *New Universities Quarterly*, vol. 35, pp. 425–38 (1981), page 426.

CHAPTER 3

1. Interview with the author, Johannesburg, February 1985.
2. "Human Evolution After Raymond Dart," in *Hominid Evolution: Past, Present and Future*, edited by Phillip V. Tobias, published by Alan Liss, New York, 1985, pp. 3–18.
3. Interview with the author, Philadelphia, 23 May 1984.
4. As 3.
5. *An Autobiography*, published by the Philosophical Library, 1950, page 480.
6. As 3.
7. As 2.

8. "Taung: A Mirror for American Anthropology," in 2, pp. 19–24.
9. "The antiquity of man," lecture to Sigma Xi, 2 December 1921, at Yale University, published by Yale University Press in *The Evolution of Man.*
10. As 2.
11. As 8.
12. "Is the Ape-Man a Myth?," in *Human Biology*, vol. 1, January 1929, pp. 4–9, page 4.
13. *Man Rises to Parnassus*, published by Princeton University Press, 1927, page 163.
14. "The Discovery of Tertiary Man," in *Science*, 3 January 1930, pp. 1–7, page 2.
15. As 14, page 7.
16. "Recent Discoveries Relating to the Origins of the Antiquity of Man," in *Science*, vol. 65, pp. 481–88, 20 May 1927, page 492.
17. Transcript, Osborn files, American Museum of Natural History, New York.
18. Transcript of seminar, 4 March 1927, Osborn files, American Museum of Natural History, New York.
19. "Two Views of the Origin of Man," in *Science*, 17 May 1927, pp. 601–5, page 602.
20. "A Short History of the Discovery and Early Study of the Australopithecines," in *Hominid Origins*, edited by Kathleen J. Reichs, published by University Press of America, 1983, page 9.
21. *Finding the Missing Link*, published by Watts and Company, 1950, page 27.
22. "A Framework of Plausibility for an Anthropological Forgery," *Anthropology*, vol. 3, pp. 47–58 (1979), page 47.

CHAPTER 4

1. In *American Anthropologist*, vol. 45, pp. 39–48 (1943), page 44.
2. "The Expulsion of the Neanderthals from Human Ancestry," in *Social Studies in Science*, vol. 12, pp. 1–36 (1982), page 5.
3. "The Fate of the Classic Neanderthals," in *Current Anthropology*, vol. 5, pp. 3–43 (1964), page 4.
4. As 2, page 20.
5. *Essays on the Evolution of Man*, published by Oxford University Press, 1924, page 41.
6. *The Earliest Englishman*, published by Watts and Co., 1948, page 103.
7. "The Poor Brain of Homo Sapiens Neanderthalensis," in *Ancestors: The Hard Evidence*, edited by Eric Delson, published by Alan R. Liss, 1985, pp. 319–24, page 319.
8. As 3, page 5.
9. As 2, page 8.
10. As 2, page 23.
11. "A Framework for the Plausibility of an Anthropological Forgery," in *Anthropology*, vol. 3, pp. 47–58 (1979), page 50.
12. "Description of the Human Skull and Mandible and the Associated Mammalian Remains," in *Quarterly Journal of the Geological Society*, vol. 69, pp. 117–47 (1913), page 139.
13. *Human History*, published by Jonathan Cape, 1934, page 85.

14. "The Dawn Man of Piltdown, Sussex," in *Natural History*, vol. 21, pp. 580–81 (1921).
15. *Fossil Men*, published by Oliver and Boyd, 1923, page 471.
16. As 11, page 51.
17. As 11, page 52.
18. As 5, page 67.
19. As 11, page 55.
20. As 13, page 84.
21. "The Controversies Concerning the Interpretation and Meaning of the Remains of the Dawn Man Found Near Piltdown," in *Memoirs and Proceedings of the Manchester Literary and Philosophical Society*, vol. 59, pp. VII–IX, 31 March 1914, page IX.
22. As 13, page 67.
23. "The Exposure of the Piltdown Fraud," lecture at the Royal Institution, London, 20 May 1955.
24. *History of the Primates*, published by the British Museum (Natural History), 1950.
25. Interview with the author, Berkeley, 3 October 1984.
26. "The Jaw of the Piltdown Man," *Smithsonian Miscellaneous Collections*, vol. 65, no. 12, 24 November 1915, pp. 1–31, page 1.
27. *Man-Apes or Ape-Men*, published by Holt, Rinehart and Winston, 1967, page 31.
28. Interview with the author, Philadelphia, 23 May 1985.
29. "The Origin of Man from a Brachiating Anthropoid," in *Science*, vol. 71, pp. 645–50 (1930), page 650.
30. "A Short History of the Discovery and Early Study of the Australopithecines," in *Hominid Origins*, edited by Kathleen J. Reichs, published by University Press of America, 1983, pp. 1–77, page 24.
31. As 27, page 23.
32. As 28.
33. Bernard Campbell "Inspiration and Controversy: Motives in Research," in *South African Journal of Science*, February 1968, pp. 60–63, page 63.
34. As 30, page 46.
35. "Myths and Methods in Anatomy," *Journal of the Royal College of Surgeons, Edinburgh*, vol. 11, pp. 87–114, page 92.

CHAPTER 5

1. "Rethinking Human Origins," in *Discovery*, vol. 13, pp. 2–9 (1978), page 9.
2. "Hominoid Evolution and Hominid Origins," in "Recent Advances in the Evolution of Primates," *Pontificiae Academiae Scripta Varia* 50 (1983), pp. 43–61, page 45.
3. "The Phyletic Position of *Ramapithecus*," *Postilla*, Yale Peabody Museum, pp. 371–76 (1961), p. 373.
4. "The Yale Fossils of Anthropoid Apes," in *American Journal of Science*, vol. 29, pp. 34–39 (1935), page 37.
5. Letter, Lewis to author, 31 October 1985.
6. Interview with the author, Duke University, 25 September 1985.
7. As 5.

8. Interview with the author, Harvard University, 23 October 1984.
9. Interview with the author, Duke University, 4 February 1986.
10. Interview with the author, New York, 13 December 1985.
11. As 3, page 374.
12. "A Source for Dental Comparison of *Ramapithecus* with *Australopithecus* and *Homo,*" in *South African Journal of Science,* February 1968, pp. 92–112, page 97.
13. As 6.
14. As 4, page 36.
15. As 12, page 97.
16. As 6.
17. As 8.
18. "Some Fallacies in the Study of Hominid Phylogeny," *Science,* vol. 141. pp. 879–89 (1963), page 879.
19. As 8.
20. As 6.
21. "An Early Miocene Member of Hominidae," in *Nature,* 14 January 1967, pp. 155–63, page 163.
22. Interview with the author, Harvard University, 14 November 1984.
23. "On the Mandible of Ramapithecus," in *Proceedings of the National Academy of Sciences,* vol. 51, pp. 528–35 (1964).
24. "Some Problems of Hominid Classification," in *American Scientist,* vol. 53, pp. 237–59 (1965), page 238.
25. "Notes on Ramapithecus, the Earliest Known Hominid, and Dryopithecus," in *American Journal of Physical Anthropology,* vol. 25, pp. 1–5 (1966), page 2.
26. "Human Origins," in *Advancement of Science,* March 1968, pp. 368–76, page 368.
27. As 8.
28. *The Descent of Man, and Selection in Relation to Sex,* published by John Murray, London, 1871, page 137.
29. As 8.
30. As 9.
31. "Maxillofacial Morphology of Miocene Hominoids from Africa and Indo-Pakistan," in *New Interpretations of Ape and Human Ancestry,* edited by R. L. Ciochon and R. S. Corruchini, published by Plenum Publishing Corporation, 1983, pp. 211–38, page 233.
32. As 26, page 377.
33. "Preliminary Revision of the Dryopithecinae," in *Folia Primatologia,* vol. 3, pp. 81–152 (1965).
34. As 8.
35. "Major Trends in Human Evolution," in *Current Argument on Early Man,* edited by Lars-Konig Konigson, published by Pergamon Press, 1978, pp. 261–85, page 266.
36. As 8.
37. As 8.
38. "*Ramapithecus* and Hominid Origins," in *Current Anthropology,* vol. 23, pp. 501–22 (1982), page 503.

39. As 8.
40. "The Early Relatives of Man," in *Scientific American*, July 1964, pp. 22–34.
41. As 6.
42. Interview with the author, British Museum (Natural History), London, 6 June 1984.
43. As 8.
44. "Reconstruction of the Dental Arcades of Ramapithecus Wickeri," in *Nature*, vol. 244, pp. 313–14 (1973).
45. As 6.
46. As 8.
47. "Adaptive Responses of Hominids to Their Environments as Ascertained by the Fossil Evidence," in *Social Biology*, vol. 19, pp. 115–27 (1972), page 117.
48. As 6.
49. "Rethinking Human Origins," in *Discovery*, vol. 13 (1), pp. 2–9 (1978).
50. As 8.
51. "Ramapithecus," in *Scientific American*, May 1967, pp. 28–35, page 28.
52. As 8.

CHAPTER 6

1. "A Molecular Approach to the Question of Human Origins," in *Background for Man*, edited by V. M. Sarich and P. J. Dolhinow, published by Little, Brown, 1971, pp. 60–61, page 76.
2. Interview with the author, Berkeley, 3 October 1984.
3. "Behavior and Human Evolution," in *Classification and Human Evolution*, published by Aldine, 1963, pp. 190–203, page 203.
4. "A Personal Perspective on Hominoid Macromolecular Systematics," in *New Interpretations of Ape and Human Ancestry*, edited by R. L. Ciochon and R. S. Corruccini, published by Plenum Press, 1983, pp. 135–150, page 138.
5. "Immunological Time Scale for Hominid Evolution," in *Science*, vol. 158, pp. 1200–1203 (1967), page 1220.
6. As 4, page 138.
7. Paper delivered at a symposium of the American Association for the Advancement of Science, Annual Meeting, Toronto, January 1981, ms., page 2.
8. Interview with the author, Berkeley, May 1981.
9. As 4, page 141.
10. As 8.
11. Interview with the author, Berkeley, 5 October 1984.
12. "The Earliest Hominids," in *Nature*, vol. 219, pp. 1335–38 (1969), page 1337.
13. "The Origin and Radiation of the Primates," in *Annals of the New York Academy of Sciences*, vol. 167, pp. 319–31 (1968), page 330.
14. "The Relationship of African Apes, Man, and Old World Monkeys," in *Proceedings of the National Academy of Sciences*, vol. 67, pp. 746–48 (1970), page 746.
15. "*Ramapithecus* and Human Origins," in *Current Anthropology*, vol. 23, pp. 501–22 (1982), page 505.
16. As 11.

17. "The Nature and Future of Physical Anthropology," in *Transactions of the New York Academy of Sciences*, vol. 32, pp. 128–38 (1960), page 129.
18. As 11.
19. As 8.
20. As 11.
21. As 4, page 145.
22. "Phyletic Divergence Dates of Hominid Primates," in *Evolution*, vol. 25, pp. 615–35 (1971), page 622.
23. As 11.
24. "The Revolution in Human Origins," in *Southwestern Anthropological Association Newsletter*, vol. XXI, no. 3, pp. 1–4 (1982), page 3.
25. Interview with author, Duke University, 4 February 1986.
26. As 11.
27. As 25.
28. As 25.
29. Interview with the author, Berkeley, 17 May 1984.
30. Interview with the author, Harvard, 23 October 1984.
31. As 25.
32. As 8.
33. "A Revision of the Turkish Miocene Hominoid Sivapithecus Meteai," in *Palaeontology*, vol. 23, pp. 85–95 (1980), page 94.
34. Interview with the author, British Museum (Natural History), London, 6 June 1984.
35. "New Hominoid Skull Material from the Miocene of Pakistan," *Nature*, vol. 295, pp. 232–34 (1982), page 234.
36. Interview with the author, Harvard, 14 November 1984.
37. As 34.
38. "Hominoid Evolution," in *Nature*, vol. 295, pp. 185–86 (1982), page 186.
39. As 34.
40. As 36.
41. Interview with the author, Duke University, 25 September 1985.
42. "Man's Immediate Forerunners," in *The Emergence of Man*, published by The Royal Society, 1981, pp. 21–41, page 34.
43. "A Reassessment of the Relationship Between Later Miocene and Subsequent Hominoidea," as in 4, page 617.
44. As 25.
45. As 36.
46. "The Descent of Hominoids and Hominids," in *Scientific American*, February 1984, pp. 84–96, page 87.
47. As 24, page 2.
48. "Rethinking Human Origins," in *Discovery*, vol. 13, pp. 2–9 (1978), pp. 5–6.
49. As 30.
50. Unpublished manuscript, 1979, page 16.
51. As 11.
52. As 41.
53. As 15, page 510.
54. As 25.
55. As 11.

56. "Hominoid Evolution and Hominid Origins," in "Recent Advances in the Evolution of Primates," *Pontificiae Academiae, Scripta Varia* 50, pp. 43–61, page 43.

57. As 36.

58. Telephone interview with author, 10 December 1986.

59. As 48, pp. 8–9.

CHAPTER 7

1. Interview with the author, Nairobi, 26 January 1985.

2. Citation in *Leakey's Luck*, by Sonia Cole, published by Collins, 1975, page 403.

3. *One Life*, published by Salem House, 1983, page 150.

4. Interview with the author, New York, 10 April 1984.

5. "Four Million Years of Humanity," lecture at the American Museum of Natural History, 9 April 1984, New York.

6. *By the Evidence*, published by Harcourt Brace Jovanovich, 1974, page 18.

7. Cited in 2, page 89.

8. *Stone Age Races of Kenya*, published by Oxford University Press, 1935.

9. Interview with the author, Duke University, 25 September 1985.

10. "Family in Search of Man," in *National Geographic*, February 1965, pp. 194–321, page 214.

11. Transcript of interview with Keith Berwick, 1969, in Leakey Foundation archives, Pasadena.

12. "The Fate of the 'Classic' Neanderthals," in *Current Anthropology*, vol. 4, pp. 3–43 (1964), page 7.

13. *The Antiquity of Man*, published by Williams and Norgate, 1915.

14. *The Antiquity of Man*, 2nd edition, 1925, page xii.

15. As 13.

16. As 14, page 340.

17. *Adam's Ancestors*, published by Methuen & Co., 1934, page 226.

18. As 17, page 203.

19. As 17, page 226.

20. As 13.

21. *Adam's Ancestors*, Torchbook edition, published by Harper & Row, 1960, page 186.

22. As 21, page 199.

23. As 21, page 221.

24. "The Chapter of Man Unfolds," in *The Year Book*, pp. 108–22 (1970), page 113.

25. "An Early Miocene Member of the Hominidae," in *Nature*, 14 January 1967, pp. 155–63, page 163.

26. As 4.

27. As 21, page 173.

28. As 21, page 180.

29. Interview with the author, Washington, D.C., 19 October 1984.

30. "A New Fossil Skull from Olduvai," in *Nature*, vol. 184, pp. 491–93 (1959).

31. As 21, page x.

32. *Illustrated London News*, September 19, 1959, pp. 288–89.

33. "Finding the World's Earliest Man," in *National Geographic,* September 1960, pp. 420–35, page 433.
34. Interview with the author, Berkeley, 5 October 1984.
35. As 33, page 434.
36. Cited in *Leakey's Luck,* by Sonia Cole, published by Collins, 1975, page 257.
37. Letter, Leakey to Le Gros Clark, 15 November 1960.
38. As 6, page 22.
39. Interview with the author, New York, 8 April 1984.
40. "Olduvai Gorge, Volume 4," mss. in press, page 50.
41. As 4.
42. As 40, page 63.
43. "A New Species of the Genus *Homo* from Olduvai Gorge," in *Nature,* vol. 202, 4 April 1964, pp. 7–9.
44. As 40, page 81.
45. "Facts Instead of Dogmas on Man's Origins," Wenner-Gren Foundation meeting, April 2–4, 1965.
46. Manuscript in Leakey Foundation archives, Pasadena.
47. Interview with the author, St. Thomas's Hospital Medical School, London, 11 June 1985.
48. "Just Another Ape," in *Discovery,* June 1964, pp. 37–38, page 37.
49. Interview with the author, South Creake, Norfolk, England, 4 June 1984.
50. Epilogue in *Adam or Ape,* edited by L. S. B. Leakey and Jack and Stephanie Prost, published by Schenkman Publishing Co., 1971.
51. Letter to *Discovery,* July 1964, page 49.
52. Letter to *Discovery,* August 1964, pp. 48–9.
53. Letter to *Discovery,* August 1964, pp. 49–50.
54. Letter to *Discovery,* October 1964, page 68.
55. As 4.
56. As 46.

CHAPTER 8

1. "Four Million Years of Humanity," lecture, 9 April 1984.
2. Interview with the author, Nairobi, 21 January 1985.
3. "In Search of Man's Past," in *National Geographic,* May 1970, pp. 712–31, page 731.
4. Interview with the author, Nairobi, 22 January 1985.
5. As 2.
6. "Early Artifacts from the Koobi Area," in *Nature,* vol. 226, pp. 228–30 (1970), page 230.
7. As 3, page 724.
8. Lecture to Leakey Foundation meeting, 25 October 1969, ms. in Leakey Foundation archives, Pasadena, page 13.
9. As 2.
10. "Further Evidence of Lower Pleistocene Hominids from East Rudolf," in *Nature,* vol. 237, pp. 264–66 (1972), page 265.
11. "Further Evidence of Lower Pleistocene Hominids from East Rudolf, North Kenya," in *Nature,* vol. 231, pp. 241–45 (1971).

12. "More Early Hominids from East Rudolf," anon., "News and Views," in *Nature*, vol. 237, pp. 250–51 (1972), page 250.

13. "Evidence for an Advanced Plio-Pleistocene Hominid from East Rudolf, Kenya," in *Nature*, vol. 242, pp. 447–50 (1973), page 450.

14. Interview with the author, Potomac, Maryland, 5 August 1984.

15. As 2.

16. As 14.

17. As 2.

18. Interview with the author, St. Thomas's Hospital Medical School, London, 11 June 1985.

19. "Remains Attributable to *Australopithecus* from East Rudolf," paper to symposium *Earliest Man and Environments in the Lake Rudolf Basin*, published by University of Chicago Press, 1967, pp. 484–89, p. 489.

20. As 2.

21. "Further Evidence of Lower Pleistocene Hominids from East Rudolf, North Kenya, 1973," in *Nature*, vol. 248, pp. 653–56 (1974), page 655.

22. "Should Fossil Hominids Be Reclassified?" anon., "News and Views," in *Nature*, vol. 248 (1974), page 635.

23. "Skull 1470," in *National Geographic*, June 1973, pp. 819–29, page 829.

24. *The Concepts of Human Evolution*, edited by Professor Lord Zuckerman, published by Academic Press, 1973, page 64.

25. As 24, page 69.

26. "Choose Your Own Ancestors," lecture at the California Institute of Technology, 1974, transcript from tape.

27. As 2.

28. Interview with the author, Nairobi, 26 January 1985.

29. Cited in *The Making of Mankind*, by Richard Leakey, published by E. P. Dutton, 1981, page 67.

30. "Hominids in Africa," in *American Scientist*, pp. 174–78 (1975), page 176.

31. "Rethinking Human Evolution," anon., "News and Views," in *Nature*, vol. 264, pp. 507–8, 9 December 1976, page 507.

32. As 30, page 176.

33. "A Systematic Reassessment of Early African Hominids," by D. C. Johanson and T. D. White, in *Science*, vol. 203, pp. 321–30 (1979), page 321.

34. As 1.

35. "Rival Anthropologists Divide on 'Pre-Human' Find," by Boyce Rensberger, *New York Times*, 18 February 1979.

36. Interview with the author, Berkeley, 17 May 1984.

37. As 36.

38. © CBS Inc. 1981. All rights reserved. Originally broadcast in May 1981 over the CBS Television Network as part of the *Universe* program series.

39. Interview with the author, Berkeley, 2 October 1984.

40. As 37.

41. Cited in "Daybreak Enquiry," by Karla Jennings, in *Express*, 30 August 1985.

42. "African Origins: A Review of the Record," in *Darwin's Legacy*, edited by Charles L. Hamrum, published by Harper & Row, 1983, pp. 25–44, page 25.

43. As 28.

44. Interview, by Helen E. Fisher, in *Omni*, March 1983, pp. 95–145, page 102.
45. Interview with the author, Harvard, 21 March 1983.
46. As 28.
47. *The Making of Mankind*, published by E. P. Dutton, 1981, page 70.

CHAPTER 9
1. Interview with the author, Birkbeck College, London, 6 December 1984.
2. Letter, Fitch to R. Leakey, 7 August 1969.
3. Interview with the author, Churchill College, Cambridge, 5 December 1984.
4. "Improvements in Potassium-Argon Dating: 1962–1975," in *World Archaeology*, vol. 7, pp. 198–209 (1975), page 202.
5. Letter, Fitch to R. Leakey, 3 September 1969.
6. Letter, R. Leakey to Fitch, 8 September 1969.
7. "New Hominid Remains and Early Artifacts from Northern Kenya," in *Nature*, vol. 226, pp. 223–24 (1970), page 223.
8. Letter, R. Leakey to Cooke, 21 November 1969.
9. *One Life*, published by Salem House, 1984, page 136.
10. Letter, Cooke to author, 30 January 1985.
11. Interview with author, Berkeley, 21 November 1985.
12. "Pliocene/Pleistocene Hominidae in Eastern Africa: Absolute and Relative Ages," in *Calibration of Hominoid Evolution*, edited by W. W. Bishop, J. A. Miller, and Sonia Cole, published by Scottish Academic Press, 1972, pp. 331–68, page 361.
13. Interview with the author, University of Utah, Salt Lake City, 12 November 1984.
14. Interview with the author, Potomac, Maryland, 5 August 1984.
15. Interview with the author, Berkeley, 3 October 1984.
16. "Evolution of Elephants and Suids in East Africa," anon., in "News and Views," in *Nature*, vol. 239, (1972), page 365.
17. Interview with the author, Los Angeles, 18 November 1985.
18. Letter, Mayr to author, 19 December 1985.
19. As 17.
20. Interview with the author, Nairobi, 22 January 1985.
21. Interview with the author, St. Thomas's Hospital Medical School, London, 11 June 1985.
22. As 13.
23. "Suidae from Plio-Pleistocene Strata of the Rudolf Basin," in *Earliest Man and Environments in the Lake Rudolf Basin*, edited by Yves Coppens, F. Clark Howell, Glynn Ll. Isaac, and Richard E. F. Leakey, published by the University of Chicago Press, 1976, pp. 251–63, page 260.
24. "Thoughts on the Workshop," as in 23, pp. 585–89, page 587.
25. Report in the *South African Journal of Science*, 19 October 1973, pp. 292–93.
26. Interview with the author, Nairobi, 23 January 1985.
27. As 3.
28. As 3.
29. Letter, Dalrymple to Isaac and Howell, 2 December 1974.
30. Review of Fitch and Miller paper, submitted to Isaac and Howell, anon.

31. Letter, McDougall to Isaac, 7 January 1975.
32. Letter, Fitch and Miller to Dalrymple, cited in Dalrymple letter to Fitch and Miller, 3 February 1975.
33. Letter, Miller to author, 5 June 1985.
34. Letter, Dalrymple to author, 21 January 1985.
35. Letter, Fitch to author, 21 March 1985.
36. As 35.
37. As 34.
38. Letter, McDougall to author, 12 September 1985.
39. As 33.
40. As 10.
41. As 15.
42. Letter, Cerling to author, 13 March 1985.
43. As 42.
44. As 3.
45. As 1.
46. Interview with the author, University of Utah, 13 November 1984.
47. As 1.
48. As 46.
49. Interview with the author, Berkeley, 30 October 1984.
50. Letter, R. Leakey to Miller, 26 November 1974.
51. Letter, Miller to R. Leakey, 6 December 1974.
52. Interview with the author, 15 October 1986.

CHAPTER 10
1. Interview with the author, 21 November 1985.
2. "East Rudolf: An Introduction to the Abundance of New Evidence," by Richard Leakey and Glynn Isaac, in *Human Origins*, edited by Glynn Isaac and Elizabeth McCown, published by W. A. Benjamin, Inc., 1976.
3. Interview with the author, Berkeley, 30 October 1984.
4. Interview with the author, Nairobi, 22 January 1985.
5. *One Life*, published by Salem House, 1984, page 168.
6. Letter, Cooke to author, 30 January 1985.
7. As 4.
8. Interview with the author, Los Angeles, 18 November 1985.
9. "Geochronological Problems and Radioisotopic Dating," in *Geological Background to Fossil Man*, edited by W. W. Bishop, published by Scottish Academic Press, 1978, pages 441–69.
10. Interview with the author, Berne, Switzerland, 14 June 1985.
11. As 10.
12. Letter, R. Leakey to Fitch, 17 May 1985.
13. Letter, Fitch to R. Leakey, 9 June 1975.
14. Letter, Fitch to Curtis, 28 May 1975.
15. Letter, Curtis to R. Leakey, 19 January 1978.
16. Letter, R. Leakey to Fitch, 16 June 1975.
17. Letter, R. Leakey to Curtis, 28 July 1975.
18. Letter, Curtis to R. Leakey, 30 August 1975.
19. Letter, R. Leakey to Cerling, 30 June 1976.

20. Interview with the author, Nairobi, 23 January 1985.
21. Interview with the author, Berkeley, 5 October 1984.
22. Interview with the author, Koobi Fora, 24 January 1985.
23. As 4.
24. As 21.
25. As 6.
26. As 8.
27. As 22.
28. As 8.
29. Interview with the author, Nairobi, 26 January 1985.
30. As 8.
31. As 29.
32. As 8.
33. Letter, White to R. Leakey, 16 September 1976.
34. Letter, R. Leakey to White, 28 September 1976.
35. Letter, White to R. Leakey, 25 May 1977.
36. As 8.
37. Interview with the author, Berkeley, 17 May 1984.
38. Letter, McDougall to author, 12 September 1985.
39. Interview with the author, University of Utah, 12 November 1984.
40. "Argon-40/Argon-39 Dating of the KBS Tuff in Koobi Fora Formation, East Rudolf, Kenya," by Frank Fitch, Paul Hooker, and John A. Miller, in *Nature*, vol. 263, pp. 740–44 (1976), page 742.
41. Interview with the author, Churchill College, Cambridge, 5 December 1984.
42. As 39.
43. Interview with the author, Los Angeles, 15 October 1986.
44. Letter, Gleadow to Hurford, 17 March 1976.
45. Letter, Gleadow to R. Leakey, 15 March 1976.
46. Letter, Gleadow to Hurford, 26 November 1976.
47. As 11.
48. Letter, Hurford to Gleadow, 8 November 1976.
49. As 10.
50. Letter, Gleadow to author, 2 September 1985.
51. Letter, Gleadow to R. Leakey, 1 December 1976.
52. Letter, R. Leakey to Gleadow, 14 December 1976.
53. As 4.
54. As 10.
55. Letter, Gleadow to Hurford, 15 February 1978.
56. As 10.
57. As 11.
58. As 10.
59. As 11.
60. As 10.
61. "Fission Track Age of the KBS Tuff and Associated Hominid Remains in Northern Kenya," in *Nature*, vol. 284, pp. 228–30 (1980), page 225.
62. As 11.
63. As 10.
64. As 11.

65. Letter, Gleadow to author, 21 March 1985.
66. As 20.
67. Interview with the author, Washington, D.C., July 1986.
68. As 20.
69. "The KBS Tuff Controversy May Be Ended," in *Nature*, vol. 284 (1980), page 401.
70. As 39.
71. "Potassium-Argon and Argon-40/Argon-39 Dating of the Hominid Bearing Sequences at Koobi Fora, Lake Turkana, Northern Kenya," in *Geological Society of America Bulletin*, vol. 96, pp. 159–75 (1985), page 161.
72. As 41.
73. As 4.
74. As 41.
75. Interview with the author, Birkbeck College, London, 6 December 1984.
76. Letter, Fitch to R. Leakey, 13 May 1981.
77. Letter, R. Leakey to Fitch, 19 May 1981.
78. Letter, McDougall to author, 12 September 1985.
79. Letter, Brown to author, 16 January 1985.
80. As 22.
81. Letter, Findlater to author, 1 February 1985.
82. As 4.
83. Interview with the author, Harvard, 14 November 1984.
84. As 4.
85. Transcript of an Australian Broadcasting Corporation program on controversies in geology, undated.
86. As 4.
87. Interview with the author, Berkeley, 21 November 1985.
88. As 87.
89. As 4.

CHAPTER 11

1. Interview with the author, Berkeley, 2 October 1984.
2. Interview with the author, Berkeley, 17 May 1984.
3. R. Leakey interview with the author, Nairobi, 26 January 1985.
4. As 1.
5. Review of *Lucy* in *Natural History*, April 1981, pp. 90–95, page 90.
6. Cited in "Daybreak Enquiry," by Karla Jennings, in *Express*, 30 August 1985, page 34.
7. "Plio-Pleistocene Hominid Discoveries in Hadar, Ethiopia," by D. C. Johanson and M. Taieb, in *Nature*, vol. 260, pp. 293–97 (1976) p. 296.
8. *Lucy* by Don Johanson and Maitland Edey, published by Simon and Schuster, 1981, pages 208–9.
9. "Rethinking Human Evolution," anon., in "News and Views," *Nature*, vol. 264, pp. 507–8, page 507 (1976).
10. Interview with the author, Nairobi, September 1977.
11. As 8, page 217.
12. As 10.
13. As 8, page 217.

14. As 8, page 218.
15. Interview with the author, Berkeley, 17 May 1984.
16. Interview with the author, Potomac, Maryland, 5 August 1984.
17. *Disclosing the Past*, published by Doubleday and Company, 1984, page 180.
18. As 15.
19. "Fossil Hominids from the Laetoli Beds," by M. D. Leakey, R. L. Hay, G. H. Curtis, R. E. Drake, M. K. Jackes and T. D. White, in *Nature*, vol. 262, pp. 460–66 (1976), page 466.
20. Interview with the author, Berkeley, 5 October 1984.
21. Interview with the author, Berkeley, 22 May 1984.
22. Interview for *The Making of Mankind*, BBC Television, 4 September 1979.
23. "Footprints in the ashes of time," in *National Geographic*, April 1979, pp. 446–57, page 446.
24. As 22.
25. As 8, page 231.
26. As 8, page 224.
27. Letter, White to M. Leakey, 25 June 1977.
28. Letter, Johanson to M. Leakey, 16 November 1977.
29. Letter, Johanson to M. Leakey, 23 December 1977.
30. Interview with the author, Paris, 21 June 1984.
31. Interview with the author, New York, 10 April 1984.
32. Letter, M. Leakey to G. Curtis, 8 October 1975.
33. As 1.
34. Letter, White to author, 4 September 1984.
35. As 1.
36. As 8, page 259.
37. As 1.

CHAPTER 12
1. Interview with the author, Berkeley, 17 May 1984.
2. Interview with the author, New York, 10 April 1984.
3. Interview with the author, Berkeley, 17 May 1984.
4. As 3.
5. Letter, White to author, 21 May 1984.
6. As 1.
7. As 2.
8. Interview with the author, Nairobi, 26 January 1985.
9. As 8.
10. Interview with the author, Berkeley, 2 October 1984.
11. As 8.
12. Interview with the author, Washington D.C., 26 October 1985.
13. As 8.
14. As 12.
15. *Disclosing the Past*, published by Doubleday and Company, 1984.
16. Interview with the author, Berkeley, 22 May 1984.
17. "Ethiopian Fossil Hominids," anon., in *Nature*, vol. 253, pp. 232–33 (1975), page 233.

18. "Difficulties in the Definition of New Hominid Species," anon., in *Nature*, vol. 278, pp. 400–401 (1979).
19. Cited in "Finding Eve's Cousin," in *Newsweek*, 29 January 1979, page 46.
20. "The Lucy Link," *Time*, 29 January 1979, page 73.
21. Cited in "The Leakey Footprints," *Science News*, pp. 196–97, February 1979.
22. Interview with the author, Berkeley, 5 October 1984.
23. As 2.
24. Interview with the author, St. Thomas's Hospital Medical School, London, 11 June 1985.
25. Interview with the author, Harvard, 24 October 1984.
26. "Tools and Tracks," in *The Emergence of Man*, joint symposium by the Royal Society and the British Academy, published 1981, pp. 95–102, page 102.
27. Letter, M. Leakey to author, 18 September 1984.
28. As 10.
29. As 22.
30. As 25.
31. Letter, Mayr to Tobias, 28 July 1981.
32. "Emergence of Man in Africa and Beyond," as in 26, pp. 43–56, page 47.
33. "Four Million Years of Humanity," lecture at American Museum of Natural History, New York, 9 April 1984.
34. Letter, Tobias to Mayr, 13 August 1981.
35. Interview with the author, Harvard, August 1981.
36. Letter, Mayr to Tobias, 14 September 1981.
37. As 26, page 102.
38. As 33.
39. Letter to *Science*, vol. 207, pp. 1102–3, 1980.
40. *Lucy* published by Simon and Schuster, 1981, page 301.
41. Interview with the author, South Creake, Norfolk, England, 4 June 1984.
42. Interview with the author, Harvard, 21 March 1983.
43. As 22.
44. As 8.
45. Cited in "The Evolution Revolution," *Chicago Sun Times*, 26 August 1979, article by Paul Galloway.

CHAPTER 13
1. "Vision with a Vengeance," in *Natural History*, September 1980, pp. 16–20, page 16.
2. "The Controversy over Human Missing Links," *Smithsonian Report* for 1928, pp. 413–65, page 413.
3. "Scientific Method and Mythological Content in Paleoanthropology," lecture given at Meeting of American Association of Physical Anthropologists, 13 April 1984.
4. As 2, page 413.
5. "Bound by the Great Chain," in *Natural History*, November 1983, pp. 20–24, page 20.
6. Cited in 5, page 24.

7. As 5, page 24.
8. Cited in 5, page 20.
9. *Lectures on Man*, London, 1864, page 128.
10. *Report of the British Association for the Advancement of Science*, pp. 144–46 (1862).
11. *Mismeasure of Man*, published by Norton, 1981, page 53.
12. *A New Theory of Human Evolution*, published by Philosophical Library, New York, page 161 (1949).
13. *Meet Your Ancestors*, published by John Long, page 11 (1948).
14. "Chapter of Conclusions," in *The Antiquity of Man*, published by Williams and Norgate, 1915.
15. *Darwinism and What It Implies*, published by Watts and Company, 1928, pp. 18–19.
16. "The Dawn Man," interview in *McClure's* magazine, vol. 55, pp. 19–28, page 27.
17. "Why Central Asia," in *Natural History*, May–June 1926, pp. 263–69, page 266.
18. *The Coming of Man*, published by Witherby, 1933, page 219.
19. "One Hundred Years of Paleoanthropology," in *American Scientist*, vol. 74, pp. 410–20 (1986), page 410.
20. *Darwinism*, published by Macmillan, London, 1889, page 469.
21. As 20, page 461.
22. As 20, page 463.
23. "The Limits of Natural Selection," in *Essays on Natural Selection*, published by Macmillan, 1871, page 359.
24. As 23, page 416.
25. As 18, page 220.
26. "The Myth of Human Evolution," in *New Universities Quarterly*, vol. 35, pp. 425–38 (1981), page 427.
27. Letter, Cartmill to author, 13 August 1986.
28. *The Descent of Man and Selection in Relation to Sex*, published by John Murray, London, 1871.
29. "Four Legs Good, Two Legs Bad," in *Natural History*, November 1983, pp. 65–78 (1983), page 68.
30. *Essays on the Evolution of Man*, published by Oxford University Press, 1924, page 40.
31. *Origins*, published by E. P. Dutton, 1977, page 208.
32. As 26, page 431.
33. As 29, page 69.
34. "Hunting: An Integrating Biobehavior System and Its Evolutionary Importance," in *Man the Hunter*, published by Aldine, 1968, pp. 304–20.
35. "The Evolution of Hunting," as 34, pp. 293–303, page 293.
36. As 29, page 77.
37. As 26, page 432.
38. As 29, page 77.
39. As 29, page 65.

AFTERWORD

1. "The Fate of the 'Classic' Neanderthals," by L. Brace, in *Current Anthropology*, vol. 5, pp. 3–43 (1964).

2. "Mitochondrial DNA Evolution and Human Evolution," by R. L. Cann et al., in *Nature*, vol. 325, pp. 31–36 (1987), page 35.

3. "The Mother of Us All—A Scientist's Story," by C. Petit, *San Francisco Chronicle*, 24 March 1986, page 1.

4. "The Myth of Eve: Molecular Biology and Human Origins," by F. J. Ayala, in *Science*, vol. 270, pp. 1930–1936 (1995), page 1930.

5. "The Search for Adam and Eve," in *Newsweek*, 11 January 1988, pp. 46–52, page 46.

6. "Argument Over a Woman," by J. Shreeve, in *Discover*, August 1990, pp. 52–59, page 52.

7. As 6, page 57.

8. As 6, page 57.

9. As 6, page 58.

10. As 6, page 58.

11. "The Demographic Modelling of Neanderthal Extinction," by E. Zubrow, in *The Human Revolution*, edited by P. Mellars and C. B. Stringer, published by Princeton University Press, 1989, pp. 212–231, page 229.

12. "African Populations and the Evolution of Human Mitochondrial DNA," by L. Vigilant et al., in *Science*, vol. 253, pp. 1503–1507 (1991), page 1507.

13. "Critics Batter Proof of an African Eve," by J. N. Wilford, *New York Times*, 19 May 1992, section C, page 1.

14. "Question of Human Origins Debated as Scientists Put to Rest the Idea of a Common African Ancestor," by H. M. Watzman, *The Chronicle of Higher Education*, 16 September 1992, page A7.

15. Cited in "Mitochondrial Eve Refuses to Die," by A. Gibbons, in *Science*, vol. 259, pp. 1249–1250 (1993), page 1249.

16. Cited in "Mitochondrial Eve: Wounded, but Not Yet Dead," by A. Gibbons, in *Science*, vol. 257, pp. 873–875 (1992), page 875.

17. "Human Origins and Analysis of Mitochondrial DNA Sequences," by B. Hedges et al., in *Science*, vol. 255, pp. 737–739 (1992), page 739.

18. "A New Approach to Studying Modern Human Origins," by M. Ruvolo, in *Molecular Phylogeny and Evolution*, vol. 5, pp. 202–219 (1996).

19. "Genetic Evidence on Modern Human Origins," by A. R. Rogers and L. B. Jorde, in *Human Biology*, vol. 67, pp. 1–36 (1995), page 30.

20. As 19, page 32.

21. *African Heritage*, by C. Stringer and R. McKie, published by Henry Holt, 1996, page 141.

INDEX